William Saville Kent

A manual of the Infusoria: including a description of all known flagellate, ciliate, and tentaculiferous Protozoa, British and foreign, and an account of the organization and affinities of the sponges

Vol. 3: Plates

William Saville Kent

A manual of the Infusoria: including a description of all known flagellate, ciliate, and tentaculiferous Protozoa, British and foreign, and an account of the organization and affinities of the sponges
Vol. 3: Plates

ISBN/EAN: 9783337220426

Printed in Europe, USA, Canada, Australia, Japan

Cover: Foto ©berggeist007 / pixelio.de

More available books at **www.hansebooks.com**

A MANUAL OF THE INFUSORIA.

VOLUME III. PLATES.

> "Our little systems have their day,
>   They have their day and cease to be ;
>   They are but broken lights of Thee,
>   And Thou, O Lord, art more than they."
>
> TENNYSON, *In Memoriam.*

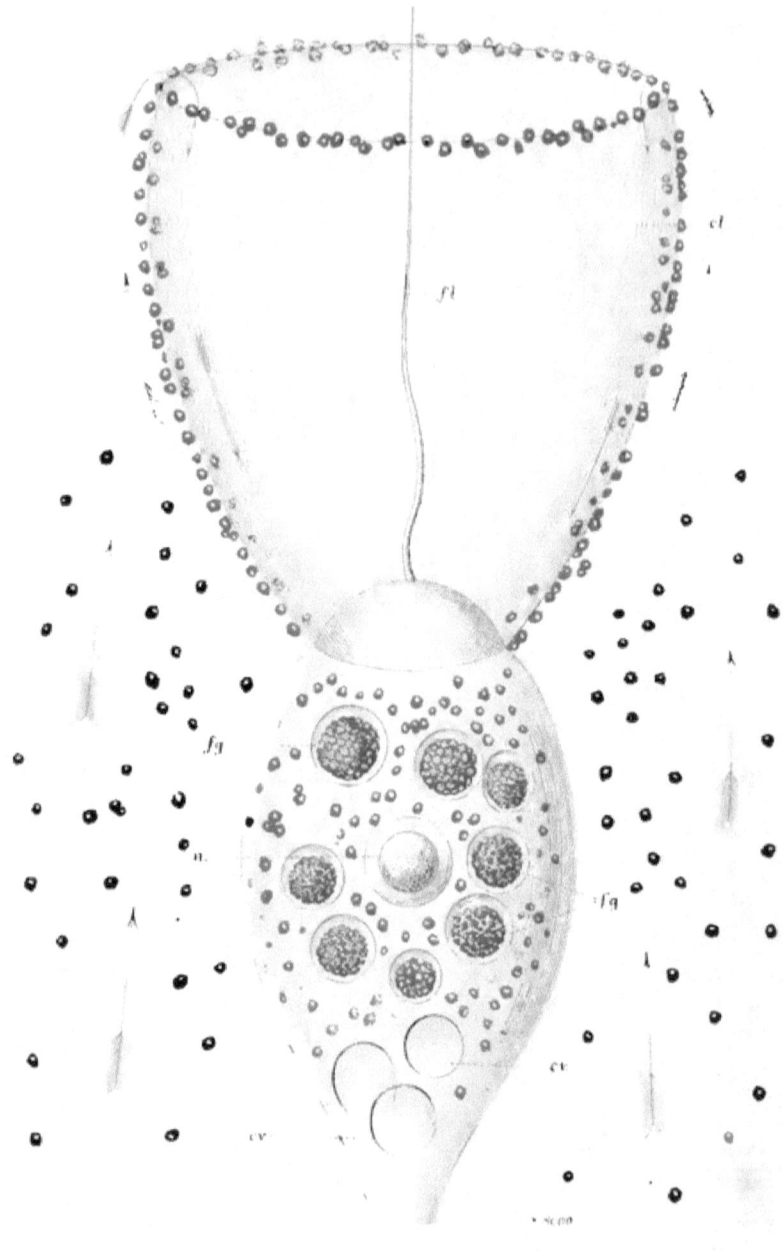

A

MANUAL OF THE INFUSORIA:

INCLUDING A DESCRIPTION OF ALL KNOWN

AGELLATE, CILIATE, AND TENTACULIFEROUS PROTOZOA,

BRITISH AND FOREIGN,

AND AN ACCOUNT OF THE

ORGANIZATION AND AFFINITIES OF THE SPONGES.

BY

W. SAVILLE KENT, F.L.S., F.Z.S., F.R.M.S.,

FORMERLY ASSISTANT IN THE NATURAL HISTORY DEPARTMENTS OF THE BRITISH MUSEUM.

VOLUME III. PLATES.

LONDON:
DAVID BOGUE, 3 ST. MARTIN'S PLACE,
TRAFALGAR SQUARE, W.C.
1880–1882.

PLATE I.

The following abbreviations retain the same significance throughout the present and succeeding Plates :—

n. Nucleus or endoplast.
cv. Contractile vesicle.
f. Flagelium.

cl. Collar.
o. Oral aperture.
an. Anal aperture.

Where the figures are borrowed from another authority, the name of such authority is bracketed; where no bracketed name appears the figures are derived from the author's original investigation.

EXPLANATION OF PLATE I.

FIG.
1, 2. TRYPANOSOMA SANGUINIS, Grube, vol. i. p. 219, × 600 (Ray Lankester).
3–6. TRYPANOSOMA EBERTHI, S. K., vol. i. p. 219, × 1200 (Eberth).
7, 8. ACTINOMONAS PUSILLA, S. K., vol. i. p. 227.—7, Zooid attached by a single stalk-like filament ; 1, zooid attached by a number of its ray-like pseudopodia, × 800.
9–11. ACTINOPHRYS SOL, Ehr., vol. i. p. 225.—Successive developmental phases out of a primary monadiform germ, as observed by the author, × 600.
12–17. MAGOSPHÆRA PLANULA, Hkl., vol. i. p. 323.—12, Adult spheroidal colony-stock, × 240 ; 13, ideal optical section of the same ; 14, a single isolated zooid derived from the disintegration of the social colony-stock ; 15, a similar zooid having assumed an amœboid phase ; 16 and 17, encysted zooids, the one at 17 having divided by segmentation into four spheroidal sporular bodies (Haeckel).
18. ACTINOMONAS MIRABILIS, S. K., vol. i. p. 227, × 800.
19, 20. MASTIGAMŒBA RAMULOSA, S. K., vol. i. p. 222.—Extended and contracted conditions, × 400.
21. MASTIGAMŒBA ASPERA, Sclz., vol. i. p. 221, × 170 (Schulze).
22, 23. MASTIGAMŒBA MONOCILIATA, Carter sp., vol. i. p. 222, dimensions unrecorded (Carter).
24. EUCHITONIA VIRCHOWII, Hkl., vol. i. p. 228, × 370 (Haeckel).
25. SPONGOCYCLIA CHARYBDEA, Hkl., vol. i. p. 229, × 72 (Haeckel).
26, 27. RHIZOMONAS VERRUCOSA, S. K., vol. i. p. 224.—26, Animalcule enclosed within granular gelatinous sheath, × 750 ; 27, example devoid of such covering.
28, 29. PODOSTOMA FILIGERUM, C. & L., vol. i. p. 225.—28, Animalcule with flagelliferous pseudopodia extended ; 29, example with appendages entirely retracted, × 250 (Clap. and Lach.).
30. MASTIGAMŒBA SIMPLEX, S. K., vol. i. p. 221.—Having attached by a posteriorly extended thread of sarcode the frustule of a diatom, probably ejected from its body, × 800.
31–33. REPTOMONAS CAUDATA, S. K., vol. i. p. 223.—31, Normal animalcule, profile view, × 800 ; 32, Dorsal view of example with short posterior pseudopodal extensions ; 33, a similar example in the act of ingesting food by the peripheral extension of its body-sarcode.
34–44. NOCTILUCA MILIARIS, Suriray, vol. i. p. 397.—34, Normal adult animalcule, × 40 ; 35 and 36, peripheral regions of two animalcules having variously developed masses of germinal bodies, × 50 (Cienkowski) ; 37, a similar germinal patch more highly magnified, and showing its composition of uniflagellate monadiform elements (Cienk.) ; 38–40, isolated monadiform germs in different aspects and phases of development, × 500 (Cienk.) ; 41, more abnormal zoospore-like germ, × 500 (Cienk.) ; 42, entire adult animalcule, dorsal view, showing median groove, stylate rod, and tooth-like process (Huxley) ; 43, latero-inferior view, showing oral cavity with tooth-like process and contained cilium (Huxley) ; 44, conjugation of two animalcules (Cienk.).
45. Vol. i. p. 399. Encysted condition of *Noctiluca*, figured by Wyville Thomson as a new diatom, *Pyrocystis pseudo-noctiluca*, × 30 (Wyv. Thom.).
46–53. LEPTODISCUS MEDUSOIDES, Hwg. (Hertwig), vol. i. p. 400.—46 and 47, Two animalcules with edges variously folded, nat. size ; 48, animalcule extended, × 40 ; 49, zooid in vertical section, showing the thicker central and more attenuate peripheral regions ; 50, segment of a similar section more highly magnified, showing oblique tubular oral fossa, superiorly attached flagellum, and at *a a a* superficial oil-like globules ; 51, endoplast or nucleus of adult animalcule, × 100 ; 52 and 53, supposed developmental phases of *Leptodiscus*, with, in the former instance, one half of the cyst-like body-wall contracted.
54. PYROCYSTIS FUSIFORMIS, Wyv. Thomson, vol. i. p. 399.—Probably the encysted condition of *Leptodiscus*, × 35 (Wyv. Thom.).

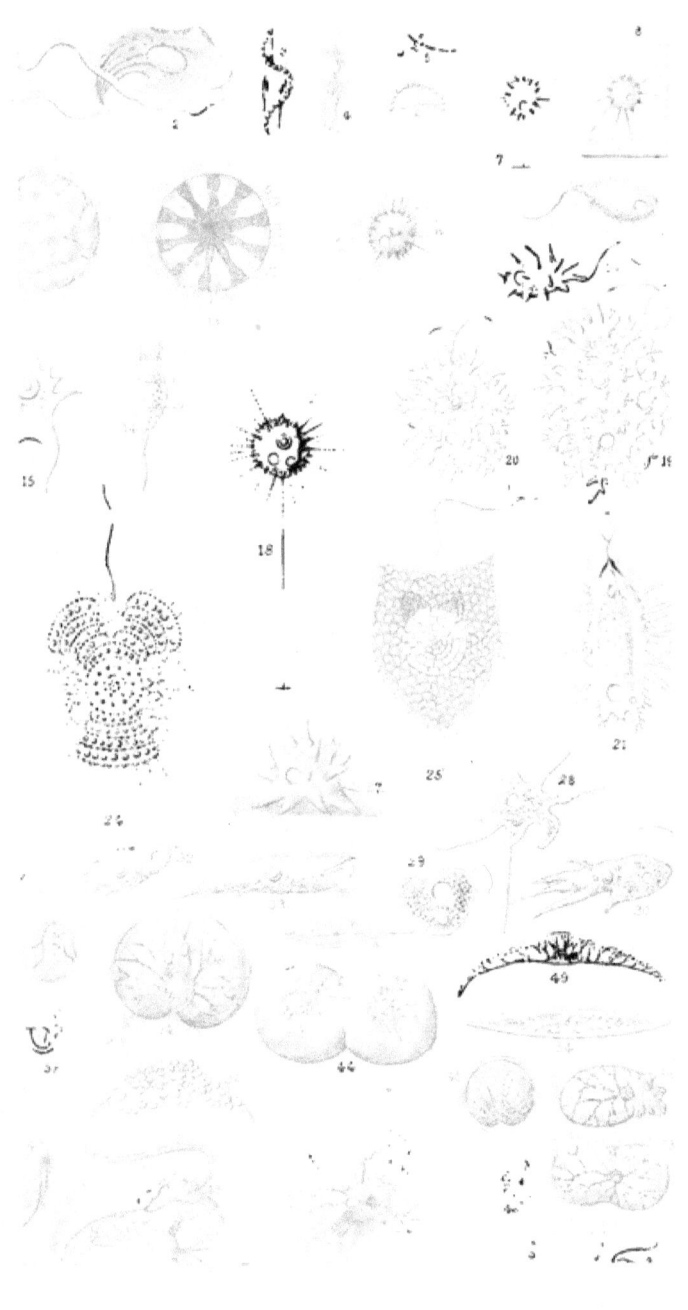

PLATE II.

EXPLANATION OF PLATE II.

FIG.

1, 2. CODOSIGA ALLIOIDES, S. K., vol. i. p. 337.—1, Umbellate adult colony-stock, or zoodendrium, bearing numerous terminal clusters of associated collared zooids, × 650; 2, a single zooid or animalcule with its body spherically, and collar conically contracted, × 800.

3. MONOSIGA GRACILIS, S. K., vol. i. p. 331, × 1200.

4–6. MONOSIGA GLOBULOSA, S. K., vol. i. p. 332.—4, Adult zooid, × 1500; 5, free-swimming monadiform germ; 6, subsequent attached condition of free-swimming germ, the characteristic collar and pedicle being as yet undeveloped.

7–9. MONOSIGA BREVIPES, S. K., vol. i. p. 332.—Exhibiting diverse protean contours, × 1200.

10, 11. CODOSIGA GROSSULARIA, S. K., vol. i. p. 338.—10, Normal adult colony-stock, × 1000; 11, smaller colony of three zooids only, having their collars conically contracted, and protruding numerous lateral, digitiform, pseudopodic processes.

12, 13. ASTROSIGA DISJUNCTA, From. sp., vol. i. p. 341.—12, Free-floating colony as imperfectly delineated by De Fromentel, the lateral margins of the collars and bases of the enclosed flagella only being represented, × 600; 13, the same colony further enlarged, the details missing in the preceding figure being added by the author.

14. CODOSIGA PYRIFORMIS, S. K., vol. i. p. 339, × 1200.

15–19. CODOSIGA FURCATA, S. K., vol. i. p. 339.—15, Colony of two zooids as observed by the author, × 1200; 16–19, imperfectly observed colony-stocks of various dimensions, as figured by Stein in the year 1854 ('Die Infusionsthiere,' Taf. iii. figs. 42 and 43), as probable young conditions of *Epistylis digitalis* or *Zoothamnium parasita*, × 450.

20–21. CODOSIGA STEINII, S. K., vol. i. p. 340, figured by Stein ('Wiegmann's Archiv,' 1849), as probable young conditions of *Epistylis* (*Opercularia*) *nutans*, × 300.

22–29. CODOSIGA (EPISTYLIS) BOTRYTIS, Ehr. sp. (*C. pulcherrima*, Jas.-Clk.), vol. i. p. 334.—22, Colony-stock with pendulous zooids diagrammatically outlined, × 1000; 23, smaller colony with three erect zooids; 24, single zooid dividing by longitudinal fission, the process having already extended through the body and the proximal region of the contracted collar; 25, two zooids assuming an amœboid condition, their collars and flagella being entirely retracted and digitiform pseudopodia protruded from all parts of their periphery; 26, a single zooid emitting similar but more slender pseudopodic processes, the collar and flagellum remaining extended, × 2000; 27, sporocyst with contained spores derived from the encystment and segmentation of a single zooid; 28, earliest illustration of the species in which the existence of the characteristic membranous collars is clearly indicated, as given by Fresenius in the year 1858; 29, associated colony-stocks crowded upon a confervoid filament, × 120.

30. DESMARELLA MONILIFORMIS, S. K., vol. i. p. 341.—A free-floating colony-stock of eight laterally united zooids, × 1200.

31, 32. MONOSIGA ANGUSTATA, S. K., vol. i. p. 330.—31, Normal adult zooid, × 2500; 32, immature or larval condition with the collar as yet undeveloped.

33–35. MONOSIGA OVATA, S. K., vol. i. p. 332.—33 and 34, Typical zooids, × 1200; 35, zooid abnormally prolonged preparatory to dividing by transverse fission.

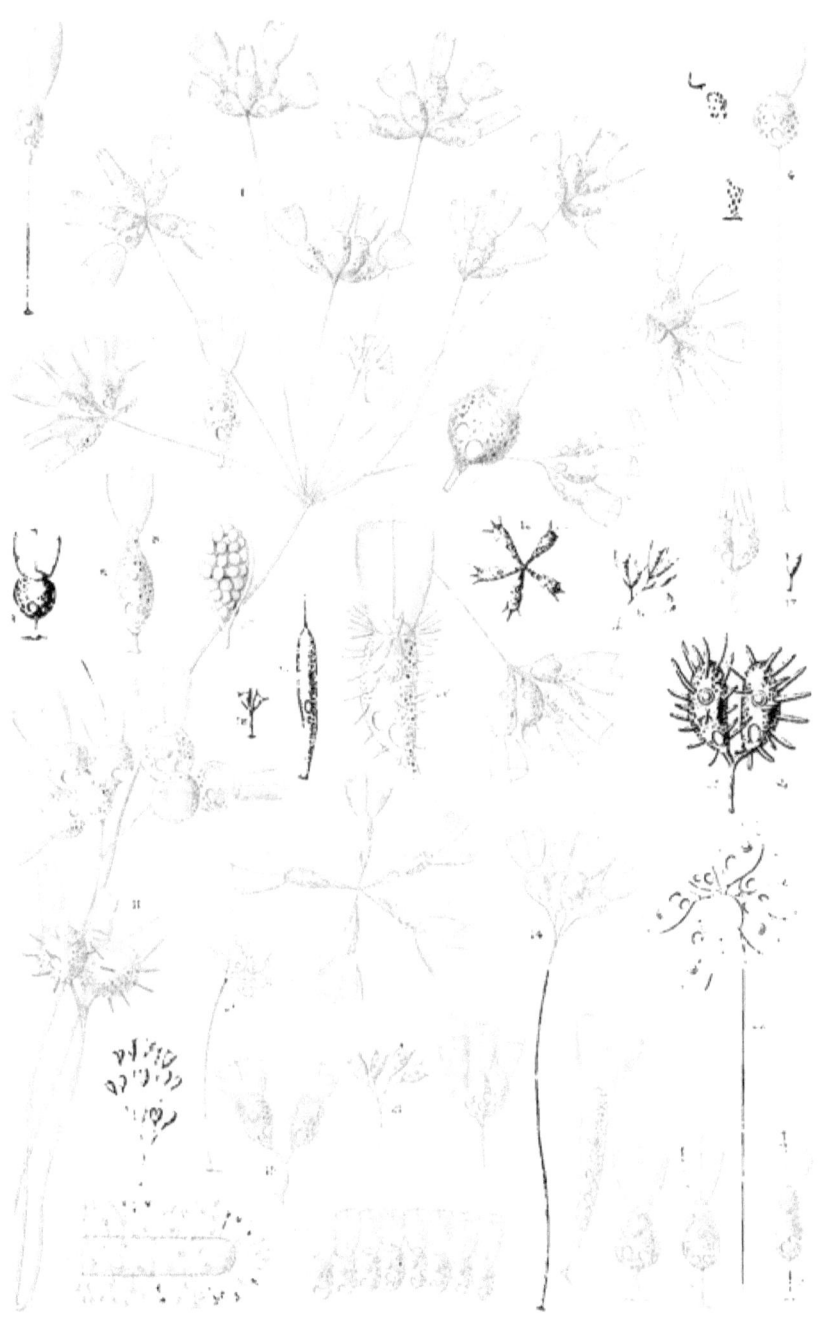

PLATE III.

EXPLANATION OF PLATE III.

Fig.
1. SALPINGŒCA or MONOSIGA sp., vol. i. p. 345.—Figured in Mr. Carter's MS. notebook without description, Bombay, Aug. 1855.

2. SALPINGŒCA or MONOSIGA sp., vol. ii. p. 703.—Figured by R. Greeff ('Wiegmann's Archiv,' Heft vi., 1870) as minute Flagellata attached to pedicle of *Epistylis flavicans*.

3-7. CODOSIGA CYMOSA, S. K., vol. i. p. 337.—3 and 4, Fully developed colony-stocks, or "zoodendria," × 1250 ; 5, branchlet with three zooids, the one at *a* being of abnormal size ; 6, branchlet with two zooids, the one at *a* having encysted and separated by segmentation into two equal halves ; 7, apparent abnormal colonial growth of the same species?

8, 9. CODOSIGA CANDELABRUM, S. K., vol. i. p. 339, × 800.

10-12. SALPINGŒCA MINUTA, S. K., vol. i. p. 347.—10, Two adult zooids and a single undeveloped germ (*a*) attached to an empty lorica of *Dinobryon sertularia*, × 1000 ; 11, a single zooid further enlarged ; 12, a zooid with collar withdrawn presenting a semi-amœboid condition.

13-15. SALPINGŒCA MARINA, J.-Clk., vol. i. p. 352.—13, Early and naked condition ; 14 and 15, adult zooids with loricæ developed, × 1800.

16. SALPINGŒCA PYXIDIUM, S. K., vol. i. p. 347, × 1000.

17-21. SALPINGŒCA AMPULLA, S. K., vol. i. p. 349.—17, Adult zooid with fully developed lorica, × 1250 ; 18, empty lorica ; 19, zooid with lorica imperfectly developed and as yet mucilaginous in consistence, at *a* a monoflagellate collarless germ attached to the exterior of the lorica ; 20, a germ which has become attached and commenced to develop its collar and protective lorica ; 21, a more advanced growth of the same zooid.

22-24. Vol. ii. p. 703. Minute flagellate and apparently collar-bearing loricate monads (*Salpingœca*), attached to pedicle of *Epistylis flavicans*, as delineated by Greeff, × 300.

25. LAGENŒCA CUSPIDATA, S. K., vol. i. p. 360, × 1500.

26. SALPINGŒCA PETIOLATA, S. K., vol. i. p. 349, × 1250.

27, 28. POLYŒCA DICHOTOMA, S. K., vol. i. p. 360.—Two social colony-stocks or polythecia, × 1000.

PLATE IV.

EXPLANATION OF PLATE IV.

FIG.

1-5. CODOSIGA UMBELLATA, Tatem sp., vol. i. p. 335.—1, More normal and adult colony-stock with compound pedicle or zoodendrium, tripartitely branched, × 625; 2, a single zooid, × 1250; 3, a simpler growth of the same fundamental formula; 4, an example with the pedicle quadripartitely branched (Tatem); 5, an abnormal type with five primary subdivisions of the supporting pedicle (Stein).

6-10. CODOSIGA (EPISTYLIS) BOTRYTIS, Ehr. sp. (Stein), vol. i. p. 334.—6, An abnormally luxuriant colony with a spheroidal cluster of associated zooids, at *a* one of the latter detached from the parent stock, × 650; 7, a colony-stock in which at *a* a larger zooid is dividing by longitudinal fission, the group of four smaller ones at *b* having been derived from similarly repeated subdivision of an original single zooid; 8, a free-swimming zooid detached from a sedentary colony; 9, coalescence or conjugation of a similar free-swimming zooid with a normal sedentary form; 10, a zooid emitting minute pseudopodic processes which present the aspect of adherent *Bacteria*.

11. SALPINGŒCA CAMPANULA, S. K., vol. i. p. 357, × 1250.

12. MONOSIGA STEINII, S. K., vol. i. p. 331.—Five zooids attached to a stalk of *Vorticella convallaria*, × 650 (Stein).

13-16. SALPINGŒCA CONVALLARIA, Stein (Stein), vol. i. p. 357.—13, Three normal zooids attached to stem of an *Epistylis*, × 650; 14, a zooid dividing by longitudinal fission; 15, zooid with lorica of an irregular and abnormal form; 16, a detached and free-swimming zooid.

17. MONOSIGA FUSIFORMIS, S. K., vol. i. p. 331.—A social group, × 1800.

18. MONOSIGA LONGICOLLIS, S. K., vol. i. p. 333, × 1800.

19-21. MONOSIGA CONSOCIATA, S. K., vol. i. p. 330.—19, A group showing at *a* a zooid, with collar and flagellum withdrawn, about to enter upon an encysted state, and to the extreme right an example with a short pedicle, × 1500; 20 and 21, zooids with collars and flagella retracted and assuming a vacuolar amœboid phase.

PLATE V.

EXPLANATION OF PLATE V.

FIG.
1-9. SALPINGŒCA AMPHORIDIUM, J.-Clark, vol. i. p. 343.—1, Social colony attached to confervoid filament, × 625; 2, a separate and normal zooid with collar fully expanded, × 1250; 3, a zooid with collar contracted, and with lorica supported on a rudimentary pedicle; 4, zooid encysted within its lorica; 5, zooid with collar entirely retracted, the flagellum remaining, but much thickened at its base, and the body-sarcode protruding in a lobose form; 6, anterior region of zooid, showing its protrusion from the lorica in the form of a fascicle of filamentous pseudopodia; 7, a zooid in which the protruded sarcode has assumed a branched, pinnatifid, contour; 8, the protruded sarcode of the same zooid having become detached, and resembling a stellate floating amœba, × 1250; 9, a more minute stellate amœba-like body found floating in the same vicinity, and probably possessing a similar derivation.

10-12. SALPINGŒCA STEINII, S. K., vol. i. p. 346.—10, two rosette-shaped colonies, and four more isolated zooids attached to the pedicle of *Epistylis anastatica*, × 300 (Stein); 11, a single zooid (Stein); 12, a colony, as found by the author, attached to the pedicle of *Vorticella campanula*, × 1000.

13. SALPINGŒCA AMPHORA, S. K., vol. i. p. 347, × 1500.

14-16. SALPINGŒCA URCEOLATA, S. K., vol. i. p. 353.—14 and 15, zooids with collar extended and contracted; 16, empty lorica, × 1500.

17-18. SALPINGŒCA RINGENS, S. K., vol. i. p. 354.—At 18 an example encysted, × 1500.

19. SALPINGŒCA CURVIPES, S. K., vol. i. p. 355, × 2000.

20. POLYŒCA DICHOTOMA, S. K., vol. i. p. 360, × 1500.

21, 22. SALPINGŒCA TINTINNABULUM, S. K., vol. i. p. 354.—22, Encysted state, × 2000.

23, 24. SALPINGŒCA (?) WALLICHII, S. K., vol. i. p. 348.—23, Remains of loricæ at *a a a* embedded within shell substance of a Globigerina; 24, a single isolated lorica, highly magnified.

25, 26. SALPINGŒCA NAPIFORMIS, S. K., vol. i. p. 355.—25, A social colony in vegetable fibre, × 800; 26, a single zooid more highly magnified.

27-31. SALPINGŒCA FUSIFORMIS, S. K., vol. i. p. 346.—27, A normal, fully expanded zooid, × 1500; 28, a zooid with collar and flagellum retracted, assuming an amœboid state; 29, an encysted zooid; 30, similar encystment, with body broken up into numerous spore-like bodies; 31, spore-like bodies further developed, and being discharged from the lorica as monoflagellate germs.

32. A form of SALPINGŒCA, apparently near S. FUSIFORMIS, dividing by transverse fission (after Bütschli).

33. A probable variety of SALPINGŒCA AMPHORIDIUM, vol. i. p. 343, the lorica having a flattened base, the collar imperfectly delineated (after Bütschli).

34. SALPINGŒCA MARINA, J.-Clk., vol. i. p. 352.—An example with the collar and flagellum withdrawn, emitting ray-like pseudopodia, × 1800.

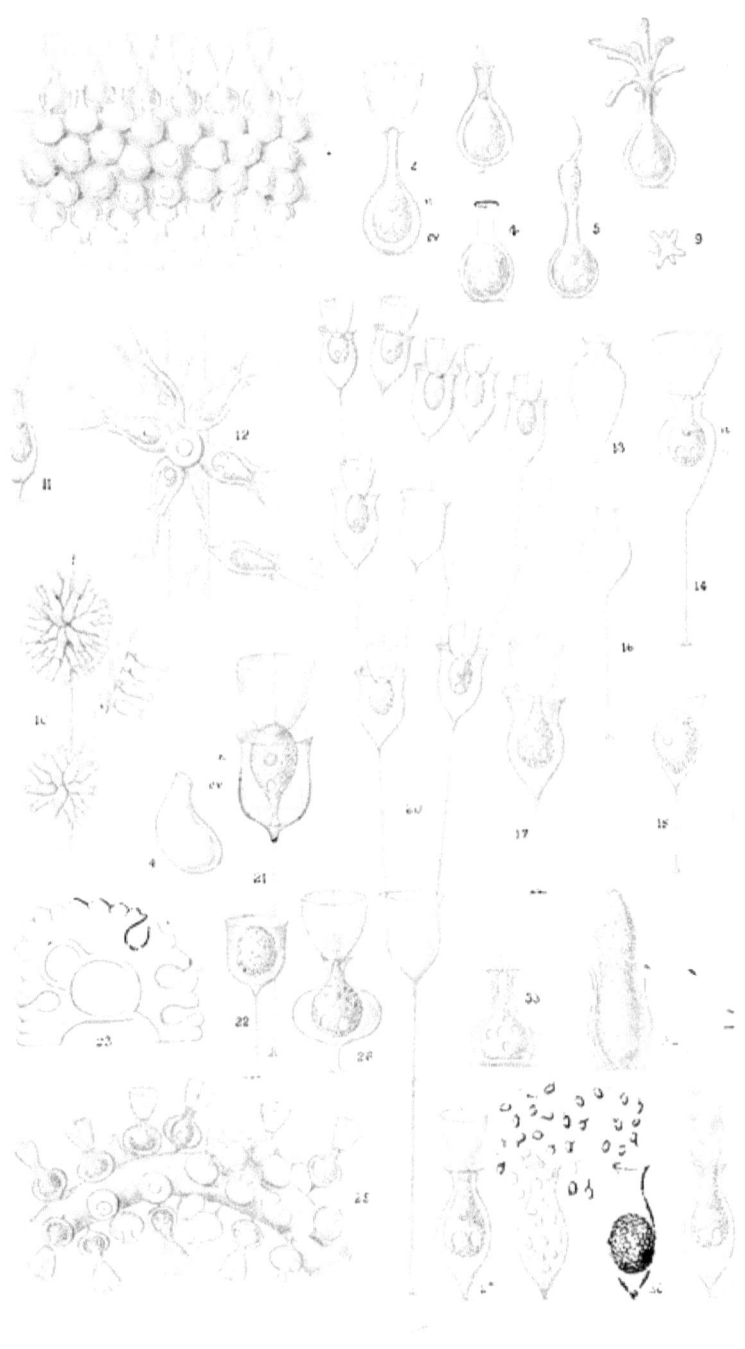

PLATE VI.

EXPLANATION OF PLATE VI.

Fig.
1-6. SALPINGŒCA INQUILLATA, S. K., vol. i. p. 354.—1, Zooid in its normal and fully extended state, × 1250; 2-5, showing various phases accompanying the process of transverse fission; 6, a recently attached collarless zooid commencing to excrete its protective lorica.

7. SALPINGŒCA LONGIPES, S. K., vol. i. p. 353.—Two zooids, × 1250.

8-16. SALPINGŒCA INFUSIONUM, S. K., vol. i. p. 356.—8. Normal adult zooid, × 800; 9, zooid dividing by transverse fission; 10, distal separated half of the same zooid presenting the form of a simple uniflagellate monad; 11, the same monadiform zooid attached by its posterior extremity, and having already developed a short pedicle; 12 and 13, further progressive phases, showing in the latter instance the zooid fully developed, but as yet wanting a lorica; 14, the same zooid, with collar and flagellum retracted, secreting its lorica; 15 and 16, sporocyst and liberated monadiform germ of the same species.

17-19. SALPINGŒCA CLARKII, Stein (after Stein), vol. i. p. 358.—17, A social colony attached to the anterior extremity of a Rotifer (*Philodina hirsuta*), × 650; 18 and 19, larger examples obtained from the roots of duckweed, the zooid in the latter instance with collar and flagella retracted and assuming an amœboid phase, × 650.

20-23. SALPINGŒCA OBLONGA, Stein (after Stein), vol. i. p. 358.—20 and 21, Normal zooids, × 650; 22, exhibiting an apparent conjugative process between a normal sedentary and a naked free-swimming zooid; 23, an encysted zooid.

24. SALPINGŒCA VAGINICOLA, Stein, vol. i. p. 352.—Apparently an intermediate variety of *S. gracilis*, J.-Clk. (after Stein), × 650.

25-32. SALPINGŒCA GRACILIS, J.-Clk., vol. i. p. 351.—25, Early condition of short-stalked variety (after Stein); 26 and 27, adult long-stalked varieties, × 1250; 28, zooid dividing by transverse fission within posteriorly pointed but non-pedicellate lorica; 29, another phase of transverse fission, the original collar and flagellum being retracted; 30, anterior half of the same zooid liberated as a free-swimming monadiform germ; 31, a zooid encysted within its lorica, and marked by a transverse divisional line; 32, a cluster of zooids having sessile and posteriorly rounded loricæ, × 1250.

33-36. SALPINGŒCA CORNUTA, S. K., vol. i. p. 350.—33, Normal adult form, the zooid adherent to the side of its lorica by several pseudopodic posterior processes, × 1000; 34, an abnormal variety, in which the second zooid derived by fission has remained closely associated with the parent one, and produced a pseudo-compound lorica; 35, an isolated zooid, having an attenuate vermicular contour; 36, a zooid attached to the wall of its lorica by a simple pedicle-like posterior prolongation.

37. SALPINGŒCA CYLINDRICA, S. K., vol. i. p. 348, × 2000.

38. SALPINGŒCA TUBA, S. K., vol. i. p. 351.—A social colony, × 1500.

39. SALPINGŒCA CARTERI, S. K., vol. i. p. 348.—Monad with so-called "ear-like points," as originally figured and described by Mr. Carter, × 1500.

PLATE VII.

EXPLANATION OF PLATE VII.

FIG.

1. HALISARCA LOBULARIS, Duj., vol. i. p. 169, a spicule-less sponge, in vertical section, after F. E. Schulze, × 75.—*af*, apertures of afferent canals or pores leading to the spheroidal monad-lined chambers or ampullaceous sacs, *amp*; *ef*, debouchment of efferent canal conducting from the ampullaceous sacs to the deeper interstitial canal-systems, which finally open upon the larger excurrent orifices or oscula ; *a, b, c, d, e, f,* progressive phases of development by segmentation of swarm-gemmules in the deeper substance of the sponge, these occupying the positions previously filled by the ampullaceous sacs, and from which, by metamorphosis, they are obviously derived ; *g*, younger and more rudimentary phase of these bodies, embedded in the cytoblastema adjacent to the ampullaceous sacs.

2. ESPERIA sp., vol. i. p. 169, a siliceous-spiculed sponge, in vertical section, showing grape-like arrangement of the ampullaceous sacs round a single afferent or pore system, × 500.—*af*, entrance of afferent canal or pore, the arrows denoting the course followed by the incurrent stream of water ; *s. cyt*, superficial cytoblastematous layer, bounding the sponge-periphery and extended canopy-wise over the distal extremities of the spicula ; *sp*, acerate spicula ; *amp*, ampullaceous sacs ; *c*, amœbiform cytoblasts ; *a b*, imperfectly developed ampullaceous sacs in the deeper substance of the cytoblastema ; 2 *a*, minute bihamate spicula from the structureless cytoblastema, × 800.

3, 4. GRANTIA COMPRESSA, Bowerbank, vol. i. p. 169, a calcareous-spiculed sponge.—3. Segment of an entire transverse section, × 300, showing at *a* central cavity or cloaca receiving currents passed through the surrounding ciliated or monad-lined chambers, *b, b* ; *c, c*, two such monad-lined chambers containing respectively one and two ciliated swarm-gemmules ; *d*, external fringe of recurvate-clavate defensive spicula. 4. One entire, and a portion of a second monad-lined chamber, × 600 ; *af*, afferent canal or pore by which currents are received from the external water ; *ef*, efferent canal conducting to the central cloacal chamber; *m*, collar-bearing monads ; *d*, external defensive spicula ; *i*, internally projecting triradiate spicula ; *g*, swarm-gemmule in its earlier amœboid and non-segmented phase of development.

PLATE VIII.

EXPLANATION OF PLATE VIII.

FIG.

1-17. GRANTIA COMPRESSA, Bowerbank, vol. i. p. 167, a calcareous-spiculed sponge.—1, Small intraspicular area, showing pavement or tesselated arrangement of collared monads; *p, p*, pore apertures, × 800; 2-8, various polymorphic forms assumed by the collared monads, × 1600; 9-12, small isolated groups of collared monads from the same sponge, exhibiting various conditions of metamorphosis, some of them with simply the collars withdrawn and flagella remaining extended, others with both these organs retracted and the body sarcode produced in form of pseudopodia, and in consequence presenting an amœbiform contour; at 9, the monads' bodies filled with ingested carmine particles; 13, an isolated monad with collar and flagellum retained in combination with a long, bifurcated, posteriorly produced pseudopodium; 15 and 16, amœbiform phases of collared monads of the same sponge; 17, a metamorphosed collared monad, with radiating pseudopodia, presenting an Actinophrys-like aspect.

18-31. HALICHONDRIA PANICEA, Johnston, vol. i. p. 167, a siliceous-spiculed sponge.—18, Group of collared monads attached to a slender acerate spicule, × 1000; 19-23, groups of collared monads attached to spicula, and in most instances partially immersed within a thin stratum of structureless cytoblastema, exhibiting various phases of metamorphosis; 24, a group of metamorphosed collared monads presenting an amœboid aspect; 25, very young, simply monoflagellate, collared monads attached to a spiculum, × 1000; 26-31, isolated amœbiform phases of collared monads of the same species, those in the last three instances having capitate pseudopodia, and presenting an Acineta-like contour.

32-40. ASCETTA PRIMORDIALIS, Hkl., vol. i. p. 167, a calcareous-spiculed sponge (after Haeckel).—32-38, Metamorphosed collared monads, × 700; 39 and 40, amœbiform zooids or cytoblasts, from the cytoblastema, the example in the latter instance possessing two nuclei or endoplasts, and representing either two recently coalesced zooids, or one about to divide by fission, × 700.

41. LEUCOSOLENIA CORIACEA, Bwbk., vol. i. p. 171.—Portions of transparent cytoblastema surrounding a poral aperture, and containing amœbiform cytoblasts enclosing ingested carmine particles, × 800.

42. Portion of cytoblastema of HALICHONDRIA PANICEA, vol. i. p. 171, containing amœbiform cytoblasts in various stages of development, those of the smallest order originating from sporular bodies, × 1500.

43. APLYSILLA SULFUREA, F. E. Sclz., vol. i. p. 171.—Associated cytoblasts, with attenuate and interconnecting pseudopodia, × 400 (after F. E. Schulze).

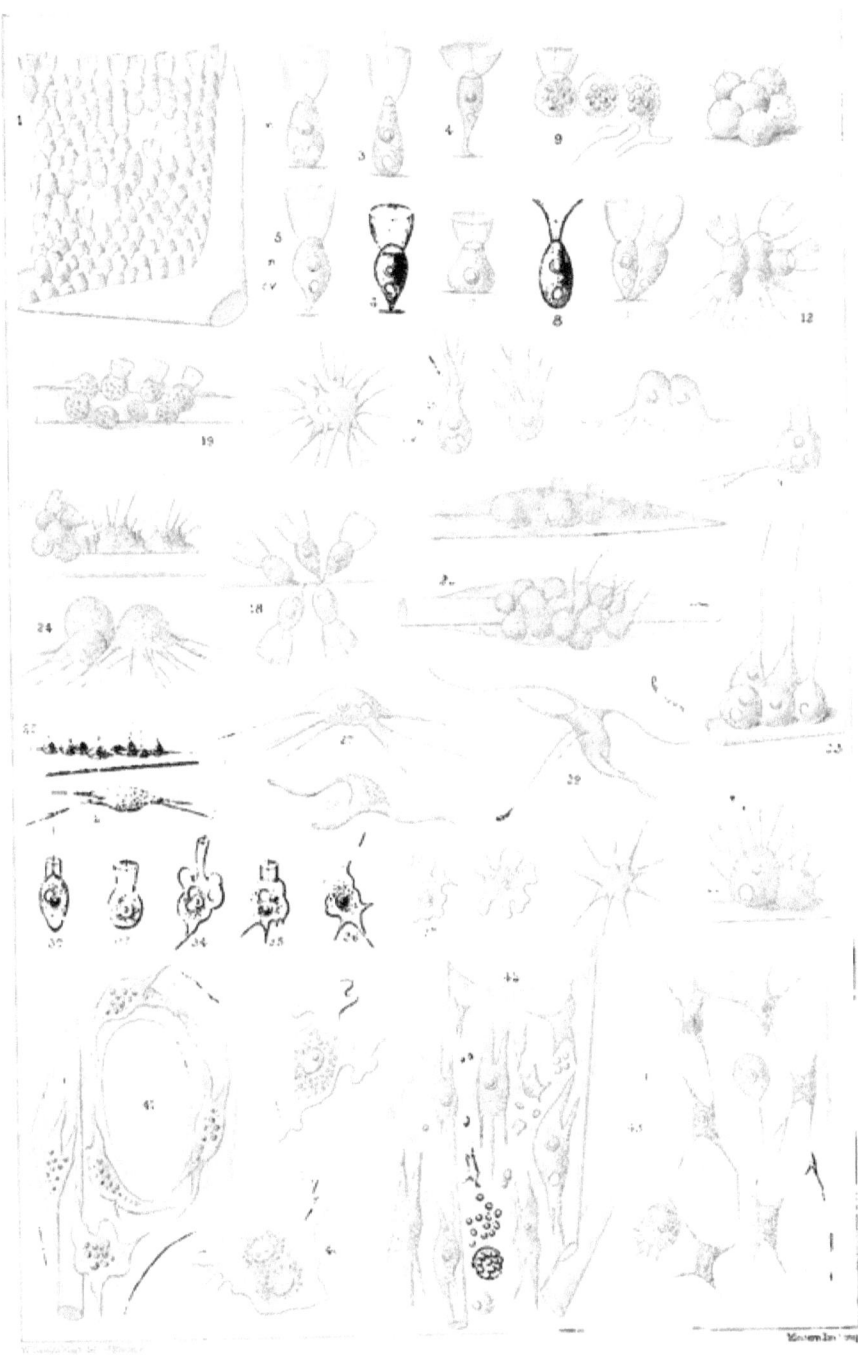

PLATE IX.

EXPLANATION.

FIG.

1–12. HALISARCA DUJARDINII, Johnston sp., vol. i. p. 188.—1, Spheroidal monad-chamber or ampullaceous sac, as seen in optical section without intersecting an afferent or efferent aperture, × 800; the introversion of this monad-chamber is alone required to produce a rosette-gemmule or ordinary swarm-gemmule as delineated at Figs. 20, 24, and 25 ; 2, six laterally attached collared monads from an ampullaceous sac of the same sponge, these corresponding remarkably in their isolated condition with the moniliform colonies of the collared monad *Desmarella moniliformis* represented at Pl. II. Fig. 30, × 1000; 3, ampullaceous sac of the same type as seen in optical section, and intersecting, where indicated by the arrows, an afferent and efferent aperture ; 4–11, progressive phases of development of an ampullaceous sac by segmentation from a primitive amœboid body, × 400 ; at 10 the segmented products present the aspect of simple amœbiform corpuscles possessing no flagellate appendages, and held together by intervening hyaline cytoblastema ; at 12 the same elements closely approximated have developed internally-projecting flagella, but still want the characteristic collars ; 12, profile view of matured ampullaceous sac, with surrounding cytoblastema ; at *c*, enclosed cytoblasts.

13–17. HALISARCA LOBULOSA, vol. i. p. 189.—13 and 14, Detached ampullaceous sac, × 400 (after Metschnikoff) ; 15–17, subspheroidal, freely detached cell-aggregations, with externally-projecting flagella, as figured and described by Metschnikoff under the title of "rosette-cells," × 400 (Metsch.).

18–21. HALISARCA DUJARDINII, vol. i. p. 190.—18–20, Spheroidal cell-combinations, or rosette-gemmules, more fully developed, as observed by the author, and shown to consist of various numerical aggregations of typical collared monads, × 800; 21, portion of the same sponge with, at *a*, earlier and undetached condition of a similar rosette-gemmule, the externally projecting units possessing flagellate appendages, but as yet no collars ; at *b*, portion of an adjacent ampullaceous sac ; *c, c*, cytoblasts immersed within surrounding cytoblastema, × 800.

22–29. GRANTIA COMPRESSA, Bwbk., vol. i. p. 178. Swarm-gemmules or so-called ciliated larvæ, exhibiting various phases and modifications of development, as observed by the author.—22 and 23, simple "planuloid" variety of such swarm-gemmule, as viewed superficially and in longitudinal optical section, and shown in the latter instance to be composed of similar closely apposed, conical, uniflagellate elements, × 350 ; 24, portion of longitudinal section of a more advanced swarm-gemmule, each constituent uniflagellate element being characterized by the possession of a distally developed rudimentary collar, which embraces

EXPLANATION OF PLATE IX. (continued).

FIG.

22–29 (continued). the base of the projecting flagellum; through the enlargement and expansion outwards of these constituent units the common body now possesses a distinctly developed central cavity; 25, more matured developmental phase of the same planuloid type of gemmule, in which the common body is shown to be composed of a symmetrically ovate aggregation of typical collared monads or spongozoa, × 600; 26–29, diverse varieties and phases of development of the "amphiblastuloid" type of swarm-gemmule from the same sponge, produced through the uneven growth of the constituent collared monads or spongozoa in the neighbouring halves of the common body; in 27, 28, and 29, the collared monads of the posterior region have developed so much in advance of those of the opposite extremity as to have withdrawn their collars and flagella, assumed an amœboid condition, and coalesced more or less completely with one another, × 600.

30. GRANTIA (SYCON) CILIATA, Bwbk., vol. i. p. 178.—Variety of "amphiblastuloid" type of swarm-gemmule, as represented by Barrois, in which an equatorial ring of metamorphosed spongozoa presents an intermediate condition of development as compared with the series above and below it.

31, 32. GRANTIA COMPRESSA, vol. i. p. 178.—Irregularly developed "amphiblastuloid" swarm-gemmules. At 31 (after O. Schmidt) the more matured amœboid units have become invaginated within the primitive central cavity of the common body.

33–35. SYCANDRA RAPHANUS, Hkl., vol. i. p. 180.—33 and 34, "Amphiblastuloid" swarm-gemmules, as represented by F. E. Schulze; in the first of these the amœboid units are invaginated within the uniflagellate, or so-called ectodermal elements, while in the second one an entire opposite process, or the invagination of the so-called ectodermal elements, is in course of progress; 35, irregularly developed amphiblastuloid swarm-gemmule (after O. Schmidt).

36–39. ASCETTA PRIMORDIALIS, Hkl., vol. i. p. 181.—36, "Planuloid" swarm-gemmule as seen in optical section, with internally contained cell-spherules; 37 and 38, portions of lateral wall of a swarm-gemmule of the same sponge, showing at *a a* cell-spherules derived by metamorphosis from the constituent uniflagellate elements; these assuming an amœboid condition and apparently coalescing with one or more neighbouring units, creep into and occupy the common central cavity, as shown in the preceding figure; 39, one such cell-spherule further enlarged, subdivided by segmentation into spore-like elements (Oscar Schmidt).

40, 41. HALISARCA LOBULARIS, Duj., vol. i. p. 189.—40, Optical transverse section of "planuloid" swarm-gemmule, showing enclosed closely corresponding and similarly derived cell-spherules, which are identified by Metschnikoff with the primitive condition of the rosette-gemmules represented at Figs. 15–17; 41, one such cell-spherule further enlarged (Metschnikoff).

PLATE X.

EXPLANATION.

Fig.

1–9. LEUCOSOLENIA CORIACEA, Bwbk., vol. i. p. 174.—1, Group of collared monads or spongozoa, certain of them at *a a* having withdrawn their collars and flagella and assumed a quiescent state, and others at *b b* become divided by segmentation into innumerable sporular elements, × 800; 2, small area of the same sponge magnified 2500 diameters, as viewed by the author with a $\frac{1}{25}$-inch objective; *a*, ordinary collared monads; *b*, one such monad with collar and flagellum withdrawn, having entered upon a quiescent or encysted state; *c*, spore-spheres, produced by the segmentation of the metamorphosed collared monads; *d d*, spores derived from the disintegration of the spore-spheres, and scattered irregularly through the common mucilaginous cytoblastema, *cyt*; *sp*, triradiate spicule; 3–7, spores in aggregate and isolated conditions liberated from the investing cytoblastema, and in most instances possessing single terminal flagella, × 2500; 8 and 9, spore-masses figured as entoderm cells by Metschnikoff, × 400.

10. ASCORTIS FRAGILIS, Hkl., showing at *a a* collared monads, and at *b* a group of sporular elements with flagellate appendages (figured by Haeckel as spermatozoa), × 400.

11. ASCETTA PRIMORDIALIS, Hkl.—A portion of the inner or lining wall; *a a*, pore-apertures circumscribed by flagelliferous monads; *b b*, spore-groups derived by metamorphosis from the ordinary collared monads, interpreted by Haeckel as sperm-cells; *c c*, large amœboid bodies derived from the metamorphosis and coalescence of similar collared cells, which develop later into the characteristic ciliated swarm-gemmules, × 350 (Haeckel).

12. Vol. i. p. 173. Spore-like bodies, found by the author associated with a species of *Halichondria*, × 600.

13. Vol. i. p. 173. Spore-like mass from the interstitial substance of a species of *Hymeniacidon*, × 500.

14, 15. LEUCOSOLENIA BOTRYOIDES, Bwbk., vol. i. p. 173.—14, An intraspicular area, consisting of a film-like expansion of transparent and structureless cytoblastema in which are immersed collared monads in an encysted state, those at *a a* having become divided by segmentation into sporular elements; at *b b*, similar spore-masses disintegrated and dispersed within the substance of the cytoblastema; *c*, a minute triradiate spiculum, developed within the cytoblastema, × 600; 15, an isolated sporocyst with contained spores from the same sponge, × 1200.

EXPLANATION OF PLATE X. (*continued*).

FIG.
16–18. Vol. i. p. 173. Sporocysts of a species of *Halichondria*, that at Fig. 18 burst and discharging minute granular spores, × 600.

19. ASCETTA PRIMORDIALIS, Hkl.—Three uniflagellate spores, or so-called spermatozoa, as figured by Ernst Haeckel, × 1600.

20–30. PROTEROSPONGIA HÆCKELI, S. K., vol. i. p. 363.—A collared monad which excretes and socially inhabits a common gelatinous matrix or "zoocytium" resembling the cytoblastema of an ordinary sponge; 20, a social colony of about forty monads, × 800; at *a a*, zooids which, withdrawing their collars and flagella, have assumed an aspect corresponding with the amœbiform cytoblasts of a sponge-body; *b b*, examples with collars and flagella retracted, dividing by transverse fission; *c c*, normal zooids, with their collars contracted; *s*, spore-mass; *z z*, hyaline mucilaginous zoocytium; 21, smaller colony of eleven zooids only, those at *a a a* exhibiting an amœbiform aspect and in one instance extending slender pseudopodia; 22, small social colony including at *a* four spore-like bodies produced by the subdivision of a metamorphosed collared zooid, and at *b b* two minute, monadiform germs; 23, small symmetrical colony-stock of sixteen zooids, derived from the even and continued segmentation of a single primary unit; 24 and 25, still younger colonies of two and four zooids only; 26, solitary attached monadiform germ, having as yet developed neither a collar nor investing zoocytium, × 1000; 27, more advanced phase of the same zooid, having a well-developed collar and mucilaginous investing sheath, and corresponding at this stage with the earlier phases of *Salpingœca ampulla*, S. K., represented at Plate III. Figs. 19 and 20; 28, metamorphosed collared zooid, projecting a lobose extension of its anterior border beyond the margin of the zoocytium; 29, spore-mass, as at 21 *s*, × 1000; 30, social colony viewed in longitudinal optical section.

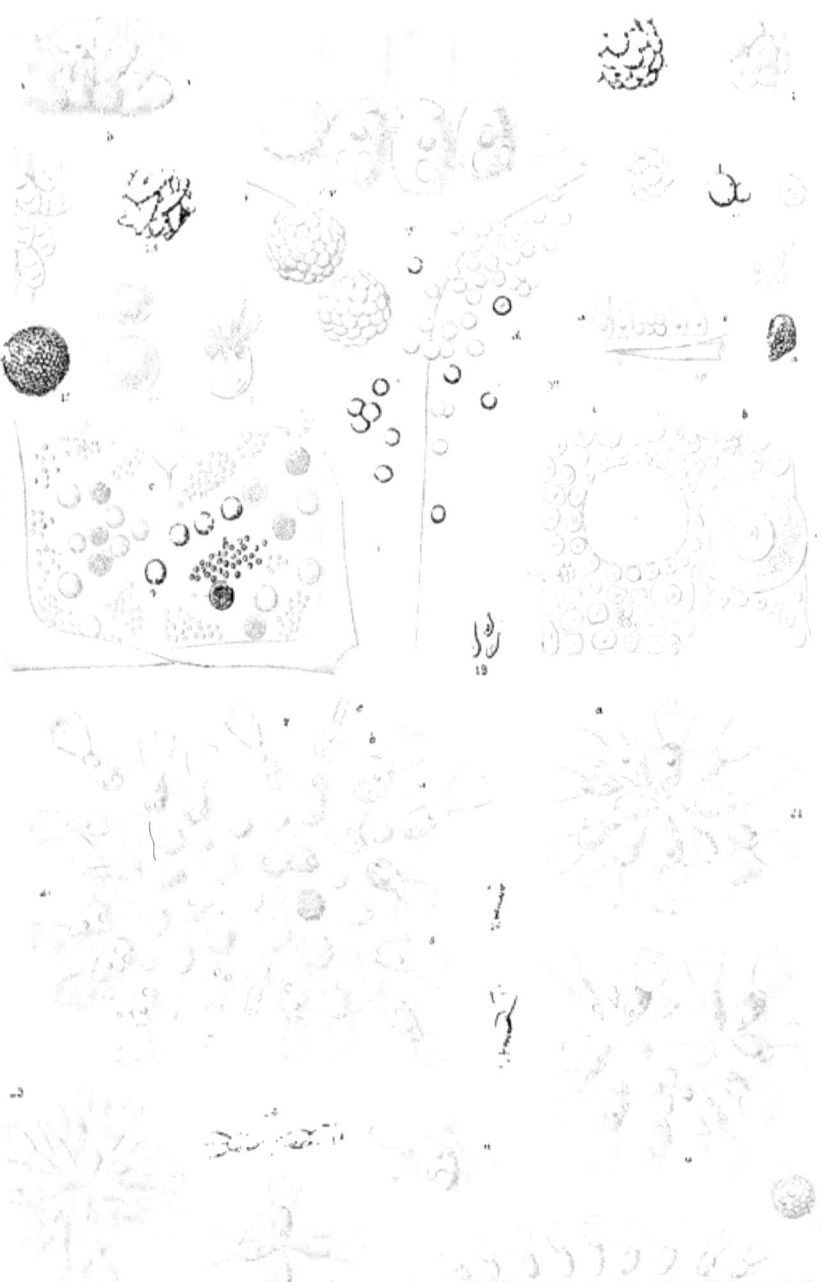

PLATE XI.

EXPLANATION.

Figs. 1–19, ILLUSTRATING DEPOSIT AND DEVELOPMENT OF INFUSORIAL GERMS ON HAY-FIBRE, AS DESCRIBED AT CHAPTER IV. PAGE 136 *et seq.*

FIG.
1. Fragment of hay-fibre examined after six hours' maceration, showing at *a a* encysted *Vorticella*, and at *b b b* masses of microspores of *Heteromita lens*, × 800.
2. Fragment of hay-fibre, wetted and immediately examined; at *a a a*, microspores of *Heteromita lens*; at *b*, four macrospores of *Oikomonas mutabilis*; at *c*, macrospores of *Heteronema caudata*; and at *d*, macrospores of an undetermined type, × 800.
3. Fragment of hay-fibre, showing at *a* a mass of microspores of *Heteromita lens* lodged in crevice formed by the serration of its surface, × 800.
4. Fragment of hay-fibre encrusted with spore-masses of the single type *Heteromita lens*.
5. Fragment of hay-fibre examined after two weeks' maceration, showing bacterial film and monad spores depending in grape-like clusters from its lower surface. Minute monads developed from these spores, mixed with vibrios and *Bacteria*, swimming beneath, × 800.

6–17. HETEROMITA LENS, Müll. sp.—6, 7, Two isolated patches of microspores, × 800; 8, 9, free-swimming monadiform germs developed from the patches of microspores, × 800; 12–15, similar monadiform germs in their primarily attached and variously aggregated free-swimming states, as seen with a $\frac{1}{12}$-inch objective, × 2500; 16 and 17, succeeding biflagellate condition of the same germs, × 800.

18. HETERONEMA CAUDATA, Duj. sp., developed from the mass of the macrospores delineated at Fig. 2 *c*, × 800.
19. HETEROMITA LENS.—Adult monad, × 800.

Figs. 20–46, ILLUSTRATING THE PROTOZOIC NATURE OF THE MYXOMYCETES OR MYCETOZOA, AS DISCUSSED AT CHAPTER II. PAGE 41 *et seq.* (FIGS. 32–35 AND 45–47 AFTER CIENKOWSKI, THE REMAINDER AFTER A. DE BARY).

20. ARCYRIA INCARNATA, Pers.—Two sporocysts or sporangia; the one at *a* burst and protruding its sporiferous rete or capillitium, × 15.
21. PHYSARUM PLUMBEUM, Fries.—Numerous associated sporangia, slightly enlarged.

EXPLANATION OF PLATE XI. (continued).

FIG.

22. ARCYRIA CINEREA, Fries.—A single sporangium, completely filled with half-matured spores, × 25.

23. PHLEBOMORPHA RUFA, de Bary.—Portion of horny rete or capillitium, with enclosed spores, × 390.

24. DIDYMIUM SERPULA, Fries.—Section of sporangium, showing contained spores.

25–28. LYCOGALA EPIDENDRON, Fries.—25, Two isolated spores; the one at *a* bursting and liberating its monadiform germ, × 390; 26, three fully developed, free-swimming, highly polymorphic monadiform zooids, developed from the spores represented in the preceding figure, × 390; 27, amœboid condition assumed by the same monadiform zooids, representing the first step towards the development of the compound plasmodium; 28, Young repent amœbiform plasmodium formed by the coalescence of several of the smaller amœboid particles.

29. ÆTHALIUM SEPTICUM, Fries.—Young repent plasmodium, × 200.

30, 31. DIDYMIUM FARINACEUM, Fries.—30, Fragment of peridium or indurated outer wall of sporangium, containing substellate calcareous spicula, × 390; 31, an isolated spicule from the same peridium further enlarged.

32–35. DIDYMIUM SERPULA, Fries.—32–34, Polymorphic monadiform zooids, the example at 32 with ingested food-particles; *n*, nucleus or endoplast; *c v*, contractile vesicle, × 350; 35, amœbiform condition of a similar monadiform zooid, × 350.

36–38. TRICHIA VARIA, Pers.—36, Isolated spore, × 390; 37 and 38, the same spore with its wall ruptured, and giving exit to a free-swimming monadiform zooid.

39. DIDYMIUM LIBERTIANUM, Fries.—Minute subaqueously developed sporangium, with nine or ten contained spores, × 390.

40–44. STEMONITIS OBTUSATA, Fries.—Successive phases of multiplication by fission of a monadiform zooid, × 390.

45–47. DIDYMIUM LEUCOPUS, Fries.—Young repent amœbiform plasmodia, containing (*c v, v c*) numerous contractile vesicles, the plasmodia at 45 and 46 being in a contracted, and at 47 in a fully extended state, × 350.

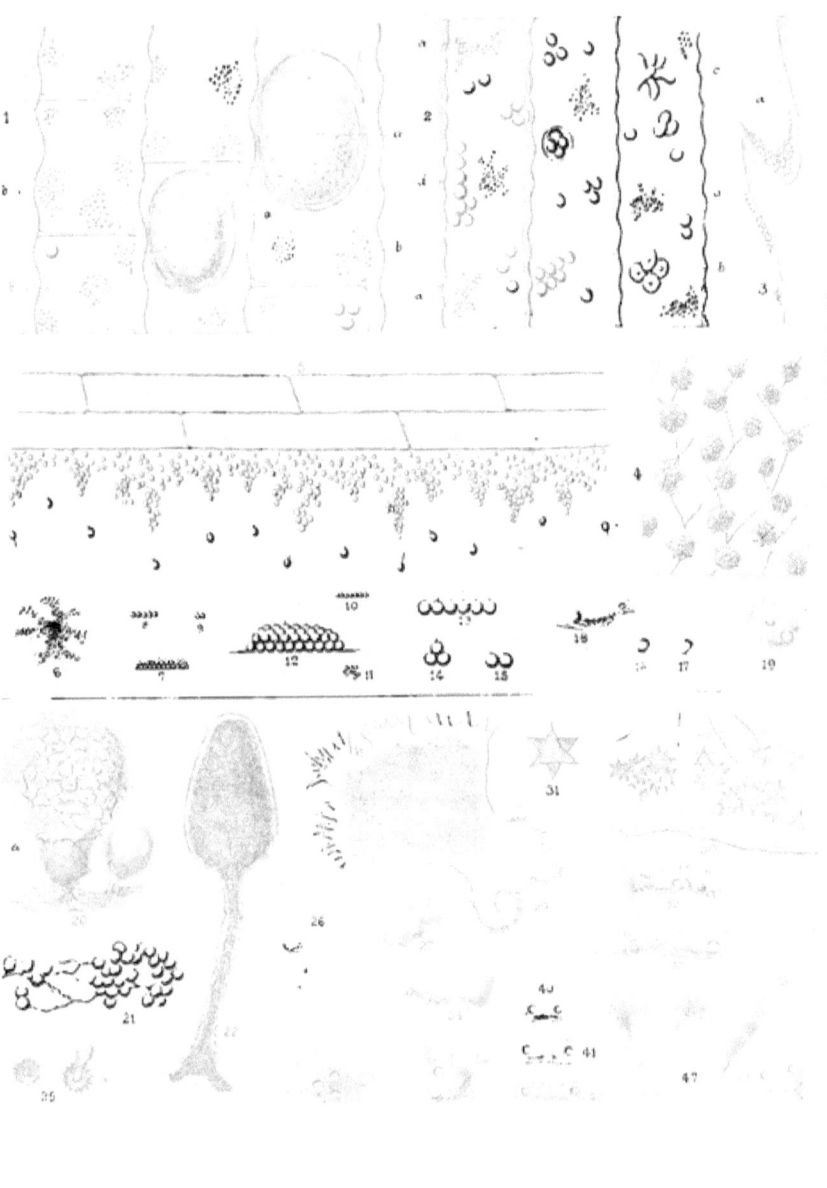

PLATE XII.

EXPLANATION OF PLATE XII.

FIG.
1-4. PHALANSTERIUM DIGITATUM, St. (after Stein), vol. i. p. 362.—1, Adult branching mucilaginous zoocytium, × 450; 2-3, early conditions of a similar zoocytium; 4, a single animalcule, × 1200; *cl*, rudimentary collar.

5-9. PHALANSTERIUM CONSOCIATUM, Cienk., vol. i. p. 362.—5, Adult discoidal colony, × 650 (after Stein); 6, an isolated animalcule further enlarged; 7 and 8, encysted zooids (after Cienkowski), the one at Fig. 8 having developed a hard tricarinate capsule; 9, an animalcule dividing by longitudinal fission.

10. SPONGOMONAS DISCUS, St., vol. i. p. 287, × 650 (after Stein).

11-14. SPONGOMONAS INTESTINALIS, Cienk. sp., vol. i. p. 287.—11, Filamentous adult colony, natural size (Cienk.); 12, extremity of similar colony enlarged, showing disposition of contained animalcules; 13, an isolated monad, × 650; 14, fragment of gelatinous granular zoocytium, showing at *a* normal zooid; *b*, two zooids derived from the longitudinal fission of such as *a*; *c*, a retracted and encysted zooid; *d, e, f*, various multiplicative phases by which a primary quiescent or encysted zooid has become divided into two, four, or eight spore-like bodies, × 650 (12-13 after Stein).

15, 16. SPONGOMONAS UVELLA, St. (Stein), vol. i. p. 288.—15, An adult colony, × 650; 16, initial condition of such a colony as founded by a single animalcule.

17-23. SPONGOMONAS SACCULUS, S. K., vol. i. p. 288.—17, Adult colony, × 10; 18 and 19, showing proportionate growth of same colony in three days as observed by the author, natural size; 20, fragment of granular zoocytium, containing two normal biflagellate and a single encysted subdividing zooid, × 1200; 21 and 22, isolated zooids, × 1200; 23, encysted zooid dividing by transverse fission.

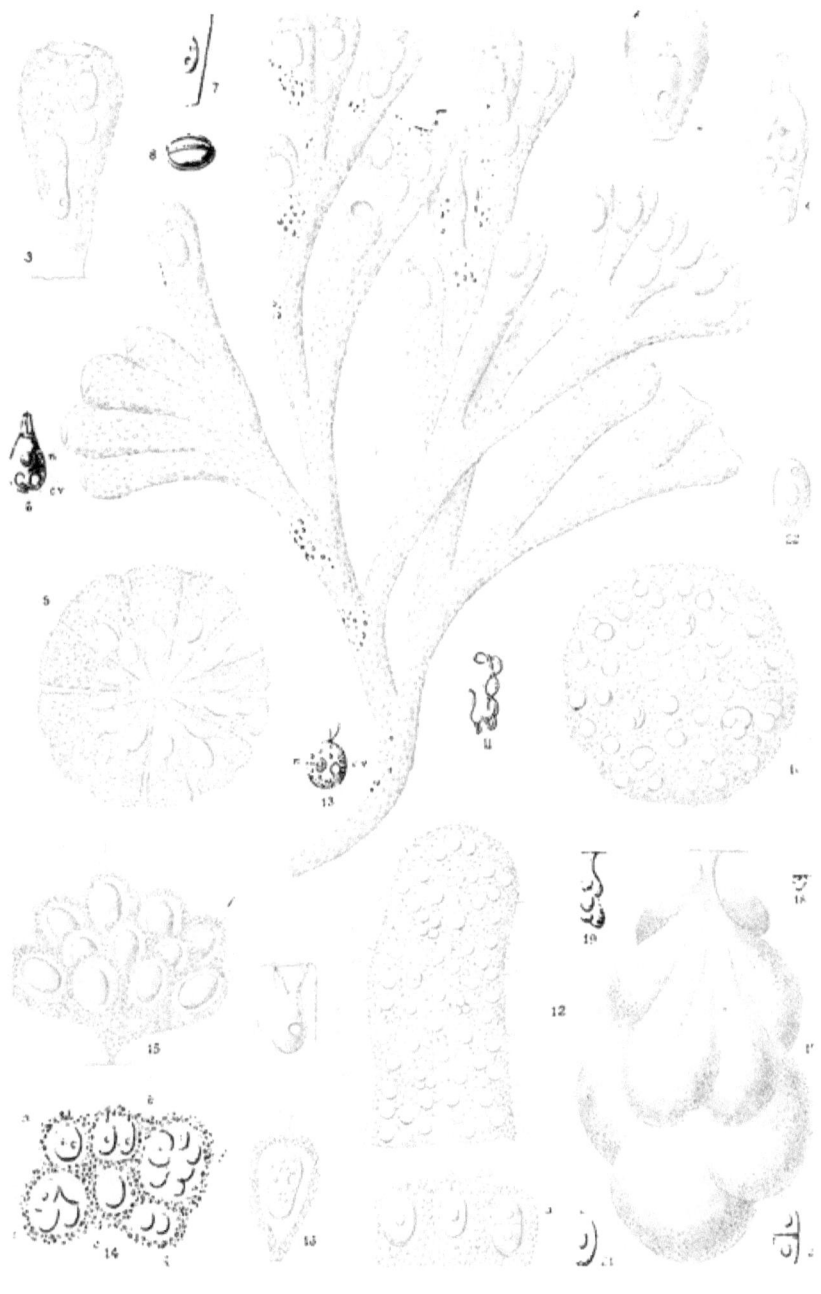

PLATE XIII.

EXPLANATION.

FIG.
1–9. MONAS DALLINGERI, S. K. (after Dallinger and Drysdale), vol. i. p. 233.—1, Normal adult form, × 2000 ; 2, monad preparing to assume an encysted state ; 3, 4, 5, progressive phases following upon encystment, and resulting in the production of a spherular aggregation of elongate vermicular macrospores ; 6, the same macrospores liberated as simple monads resembling the parent, but of smaller size ; 7, conjugation of larger and smaller monads ; 8, encystment resulting from such conjugation ; 9, the compound cyst bursting and liberating infinitesimally minute microspores.

10–18. MONAS FLUIDA, Duj., vol. i. p. 234.—10, Typical adult monad, × 1000 ; 11–16, metamorphic forms of similar adult monads ; 17, conjugation of two zooids ; 18, their encystment.

19. MONAS IRREGULARIS, Pty. (Cienkowski), vol. i. p. 236, × 350.

20, 21. MONAS OBESA, Stein sp., vol. i. p. 236, × 650 (Stein).

22–24. MONAS RAMULOSA, Stein sp., vol. i. p. 235, × 600 (Stein).

25, 26. LEPTOMONAS BUETSCHLII, S. K. (Bütsch.), vol. i. p. 243.—25, A group attached by their posterior extremities, × 600 ; 26, a free-swimming monad, × 1500.

27, 28. OPHIDOMONAS JENENSIS, Ehr., vol. i. p. 244, × 600 (Ehr.).

29–34. HERPETOMONAS MUSCÆ-DOMESTICÆ, Burnet sp. (Stein), vol. i. p. 245.—29–32, Polymorphic phases of the adult monads, × 650 ; 33 and 34, monads dividing by longitudinal fission.

35–40. HERPETOMONAS LEWISII, S. K. (Lewis), vol. i. p. 245.—35–38, Various contours of the adult organism, × 800 ; 39, two red and forty-one colourless corpuscles of the rat's blood which they inhabited, equally magnified to show proportionate size.

41, 42. SCYTOMONAS PUSILLA, St., vol. i. p. 241, × 650 (Stein).

43, 44. PLEUROMONAS JACULANS, Pty., vol. i. p. 249, × 500 (Perty).

45. MEROTRICHA BACILLATA, Meresch., vol. i. p. 249 (Dimensions unrecorded) (Mereschkowski).

46, 47. CYATHOMONAS ELONGATA, From., vol. i. p. 243.—47, Dividing by longitudinal fission, × 600 (De Fromentel).

48. CYATHOMONAS TURBINATA, From., vol. i. p. 242, × 400 (De Fromentel).

EXPLANATION OF PLATE XIII. (*continued*).

FIG.
49-53. ANCYROMONAS SIGMOIDES, S. K., vol. i. p. 247.—49, Free-swimming animalcule, × 1500; 50, animalcule fixed by distal termination of the single flagellum, the dotted outline indicating the position alternately assumed by the body with relation to the flagellum in the course of its rapid oscillations, × 2500; 51-53, showing the several progressive phases of oblique fission.

54. PLATYTHECA MICROPORA, St., vol. i. p. 262.—Three loricæ with their contained animalcules attached to a joint of conferva; at *a* two zooids, the result of fission, occupy the same lorica, × 650 (Stein).

55-64. OIKOMONAS MUTABILIS, S. K., vol. i. p. 250.—55. A group of four monads attached to vegetable fibre, showing at *a* and *b* normal sedentary forms, at *c* an example ingesting food-matter at its lateral periphery, and at *d* a young and recently adherent example, not having yet developed a filiform pedicle, × 800; 56, an adult monad about to exchange its sedentary for a free-swimming condition; 57, the same monad detached and free-swimming, still retaining an attenuation of its posterior and previously fixed extremity; 58, typical free-swimming zooid; 59 and 60, more aberrant forms; 61, a motile zooid dividing by longitudinal fission; 62 and 63, spore-masses produced by segmentation of encysted animalcules; 64, a young monad developed from a spore.

65-70. OIKOMONAS STEINII (S. K.), vol. i. p. 253.—65, A group of monads attached to a spheroidal bacterial mass, × 650; 66, four monads similarly attached, exhibiting a considerable irregularity of contour; 67, a free-swimming animalcule with branched posterior extremity; 68, a free-swimming animalcule dividing by transverse fission; 69 and 70, young free-swimming monads (Stein).

71. OIKOMONAS QUADRATA, S. K., vol. i. p. 254.—A group of five monads; at *a* and *b*, zooids ingesting food at opposite regions of the periphery. × 800.

72. OIKOMONAS OBLIQUA, S. K., vol. i. p. 252, filled with artificially administered carmine-particles, a portion of which is discharging from its posterior extremity, × 2500.

73-77. OIKOMONAS ROSTRATA, S. K., vol. i. p. 253.—73, A group of monads attached to a vegetable-fibre, × 800; 74 and 75, two attached monads, the one with and the other without a posteriorly developed pedicle, × 1000; 76, an example with an abnormally long pedicle; 77, a free-swimming monad with pedicle retracted.

78-80. OIKOMONAS TERMO, J.-Clark sp., vol. i. p. 251.—78, A group attached to vegetable fibre, showing at *a* a zooid dividing by longitudinal fission, at *b* an example ingesting food-matter, and at *c* a young, recently attached and almost stalkless zooid, × 1000; 79, a free-swimming monad; 80, a free-swimming zooid dividing by fission.

PLATE XIV.

EXPLANATION OF PLATE XIV.

FIG.
1–3. BODO JULIDIS, Leidy, vol. i. p. 256, × 750 (Leidy).
4–6. BODO MAXIMUS, Schmarda, vol. i. p. 257, × 300 (Schm.).
7–8. BODO URINARIUS, Hassall, vol. i. p. 258, × 400 (Hass.).
9–11. BODO LYMNÆI, Stiebel sp., vol. i. p. 257.—9 and 10, Motile zooids, × 600; 11, sporocyst with escaping germs, × 250 (Ecker).
12, 13. BODO (CRYPTOBIA) HELICIS, Leidy, vol. i. p. 256, × 400 (Leidy).
14. BODO INTESTINALIS, Ehr., vol. i. p. 255, × 400.
15, 16. CERCOMONAS CRASSICAUDA, St., vol. i. p. 260.—At 16, with posterior irregular pseudopodic extensions, × 600 (Stein).
17–20. CERCOMONAS LONGICAUDA, Duj., vol. i. p. 259.—17 lateral, 18 dorsal view; 19 and 20, progressive stages of longitudinal fission, × 600 (Stein).
21. CERCOMONAS CYLINDRICA, Duj., vol. i. p. 260, × 1200 (Dujardin).
22–30. CERCOMONAS TYPICA, S. K. (Dallinger and Drysdale), vol. i. p. 259.—22, Normal adult monad, × 1750; 23 and 24, amœboid phases of matured monads; 25 and 26, successive results of coalescence of two amœboid monads; 27, sporocyst ultimately derived from foregoing coalescence; 28, sporocyst burst and liberating minute spores; 29 and 30, progressive phases of transverse fission.
31–33. GONIOMONAS TRUNCATA, Fres. sp., vol. i. p. 280.—At *a*, eye-like pigment-band, × 600 (Stein).
34–36. SPUMELLA VIVIPARA, Ehr. sp. (Stein), vol. i. p. 306.—34 Attached, 35 and 36 free-swimming conditions; at *a*, eye-like pigment-band or supposed oral aperture, × 600.
37–45. PHYSOMONAS SOCIALIS, S. K., vol. i. p. 263.—37, A group of five monads attached to vegetable fibre, showing at *a* and *b* examples incepting food-matter at opposite regions of their periphery, × 1000; 38, a free-swimming monad; 39 and 40, illustrating the alternating systole and diastole of the two medianly located contractile vesicles; 41 and 42, phases of longitudinal fission; 43, an encysted group, showing at *a* a stalked and at *b* a stalkless cyst; at *c*, two stalks connected with a single and larger cyst, indicating its derivation from the conjugation or coalescence of two zooids, × 1500; 44, a sporocyst with ripe spores, × 2000; 45, monadiform germs released from the same sporocyst, × 2500.
46–52. SPUMELLA GUTTULA, Ehr. sp. (Stein), vol. i. p. 305, × 600.—46, Normal attached monad; 47 and 48, free-swimming monads; 49 and 50, illustrating conjugative process of larger and smaller monads; 51 and 52, successive phases of longitudinal fission.
53. CODONŒCA COSTATA, J.-Clk., vol. i. p. 261, × 1000 (J.-Clark).
54. CODONŒCA OBLIQUA, S. K., vol. i. p. 261, × 800.
55–59. AMPHIMONAS GLOBOSA, S. K., vol. i. p. 281.—55, A group of four monads, one, at *a*, incepting food on its lateral periphery, × 800; 56–59, successive phenomena observed during the inception of a large Bacillus, and in the preliminary phases of which process (Figs. 56 and 57) a film-like expansion of sarcode was extended over the captured prey.
60–65. DELTOMONAS CYCLOPUM, S. K., vol. i. p. 283.—60, A social group attached to hair of a species of *Cyclops*; at *a*, a free-swimming animalcule, and at *b* a young non-flagellate germ, × 1500; 61, a single monad, × 3000; 62, a group of four monads united by their posterior extremities; 63, longitudinal fission; 64 and 65, sporocysts with spores.
66. AMPHIMONAS DIVARICANS, S. K., vol. i. p. 282, × 2500.

PLATE XV.

EXPLANATION OF PLATE XV.

FIG.

1–17. HETEROMITA LENS, Müll. sp., vol. i. p. 291.—1 and 2, normal adult monad in the fixed and free-swimming conditions, × 800; 3 and 4, irregular-shaped amœbiform conditions; 5 and 6, two monads attached close to each other and about to coalesce; 7, coalescence or conjugation; 8 and 9, phases succeeding conjugation, productive in the last instance of a spheroidal sporocyst; 10, sporocyst, with contents consisting of innumerable microspores; 11–14, minute, uniflagellate, monadiform germs, developed from such microspores, and either single or adherent in diverse combinations, × 2500; 15, two sporocysts, containing eight or sixteen macrospores; 15, a sporocyst with four macrospores only; 17, the same sporocyst burst open and giving birth to four biflagellate germs differing only in size from the parent animalcule, × 1000.

18–28. HETEROMITA ROSTRATA, S. K. (Dallinger and Drysdale), vol. i. p. 293.—18, attached condition of adult monad, the outlines at a and b indicating the positions to which the body is projected by the uncoiling and extension of the posterior and adherent flagellum, × 1500; 19 and 20, successive phases of longitudinal fission; 21 and 22, conjugation of two monads; 23, triangular sporocyst resulting from such conjugation; 24, the same sporocyst bursting at its angles and releasing the enclosed microspores, × 2000; 25 and 26, development at phases of the released microspores, liberated from the triangular sporocyst, × 2500; 27 and 28, further progressive stages towards the attainment of the parent form at the end of eight and ten hours, × 2500.

29–41. HETEROMITA UNCINATA, S. K. (Dall. and Drysd.), vol. i. p. 294.—Normal adult monad, × 1500; 30, irregular semi-amœboid phase, preceding fission; 31, fission; 32, conjugation of two monads of diverse size; 33, conjugation of four monads; 34–39, sporocyst resulting from conjugation with successive phases of segmentation of its contents, producing finally a mass of minute microspores; 40, bursting of sporocyst and release of contents as minute uniflagellate germs; 41, one such germ a few hours later, having nearly attained the parent form.

42, 43. PSEUDOSPORA VOLVOCIS, Cienk., vol. i. p. 304, × 400.—Natatory and repent conditions of the same zooid (Cienkowski).

44. HETEROMITA ADUNCA, Meresch., vol. i. p. 297, × 1500 (Meresch.).

45, 46. COLPONEMA LOXODES, St., vol. i. p. 297.—45, dorsal, and 46, ventral aspect, × 600 (Stein).

47, 48. PHYLLOMITUS UNDULANS, St., vol. i. p. 299, × 600 (Stein).

49–60. SPIROMONAS ANGUSTATA, Duj. sp., vol. i. p. 298.—49–52, Free-swimming monads, with their bodies in all cases, excepting 52, variously contorted, × 1500; 53, an adherent monad; 54, amœbiform condition; 55, two amœbiform monads conjugating; 56–59, successive phases of encystment and spore-production; 60, free-swimming monad, as represented by Dujardin.

61–64. HETEROMITA GLOBOSA, St. sp. (Stein), vol. i. p. 295.—61–63, Normal adult monads, × 600; 64, a number of monads tenanting the cell of an Œdogonium, and feeding on its contents.

65, 66. HETEROMITA OVATA, Duj., vol. i. p. 295.—Dorsal and lateral aspects, × 600 (Stein).

67–78. POLYTOMA UVELLA, Müll. sp., vol. i. p. 302.—67–69, Animalcules showing looplike flexure of the basal region of the flagella, the example in the first instance being attached by this loop-like coil, as observed by the author, × 800; 70 and 71, examples with supposed basal inflations of the flagella, or independent knobbed appendages, as represented by Messrs. Dallinger and Drysdale; 72–74, successive phases of multiplication by complete segmentation of the internal contents (Dallinger and Drysdale); 75, multiplication by the breaking up of the endoplasm of the posterior region only into minute angular germs (D. and D.); 76, the same angular germs, more highly magnified (D. and D.); 77, conjugation of two animalcules (D. and D.); 78, rupture of sporocyst resulting from such conjugation, and release of microspores, × 800 (D. and D.).

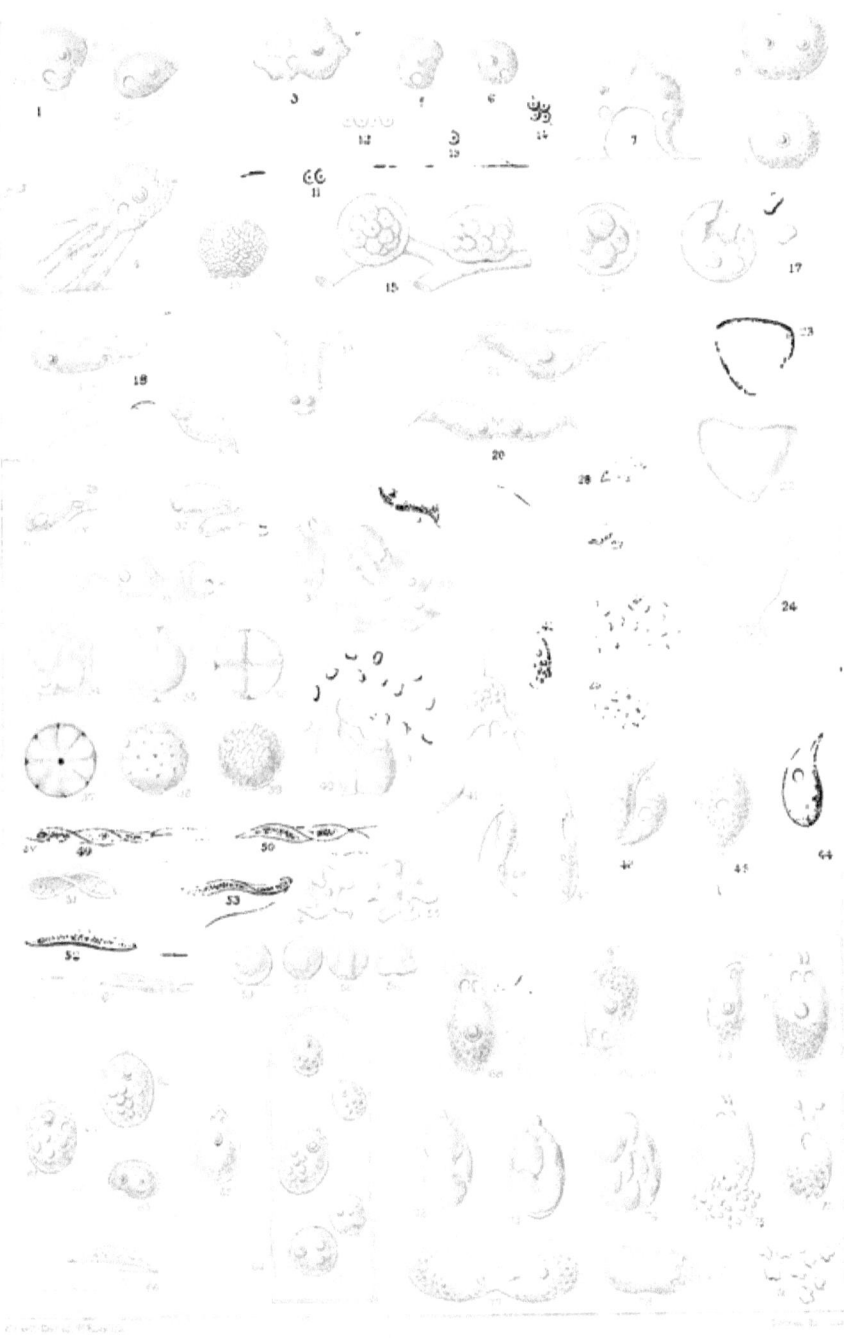

PLATE XVI.

EXPLANATION OF PLATE XVI.

FIG.

1–3. RHIPIDODENDRON SPLENDIDUM, St. (Stein), vol. i. p. 285.—Flabelliform compound colony-stock or zoothecium, consisting of several hundred closely-united tubules, the entire structure representing the product by repeated subdivision and excretion of a primarily single monadiform animalcule, × 400; 2, young zoothecium, constructed up to the point of bifurcation by a single monad, which then dividing by longitudinal fission, has produced two parallel tubes, × 600; 3, an isolated monad, × 1150.

4–8. RHIPIDODENDRON HUXLEYI, S. K., vol. i. p. 286.—4, Detached ramuscule from an adult branching zoothecium, × 170; 5, portion of branchlet, showing quadruple tubular construction, × 300; 6, distal termination still more magnified, showing contained monads, × 800; 7, diagrammatic illustration of mode in which the characteristic branching zoothecium is produced, the four primary monads, as at a, divided by longitudinal fission, the right and left halves of such total product then parting, as at b, and forming their respective tubules at divergent angles; 8, symmetrically developed adult zoothecium, × 5; 9, an adult zoothecium, natural size.

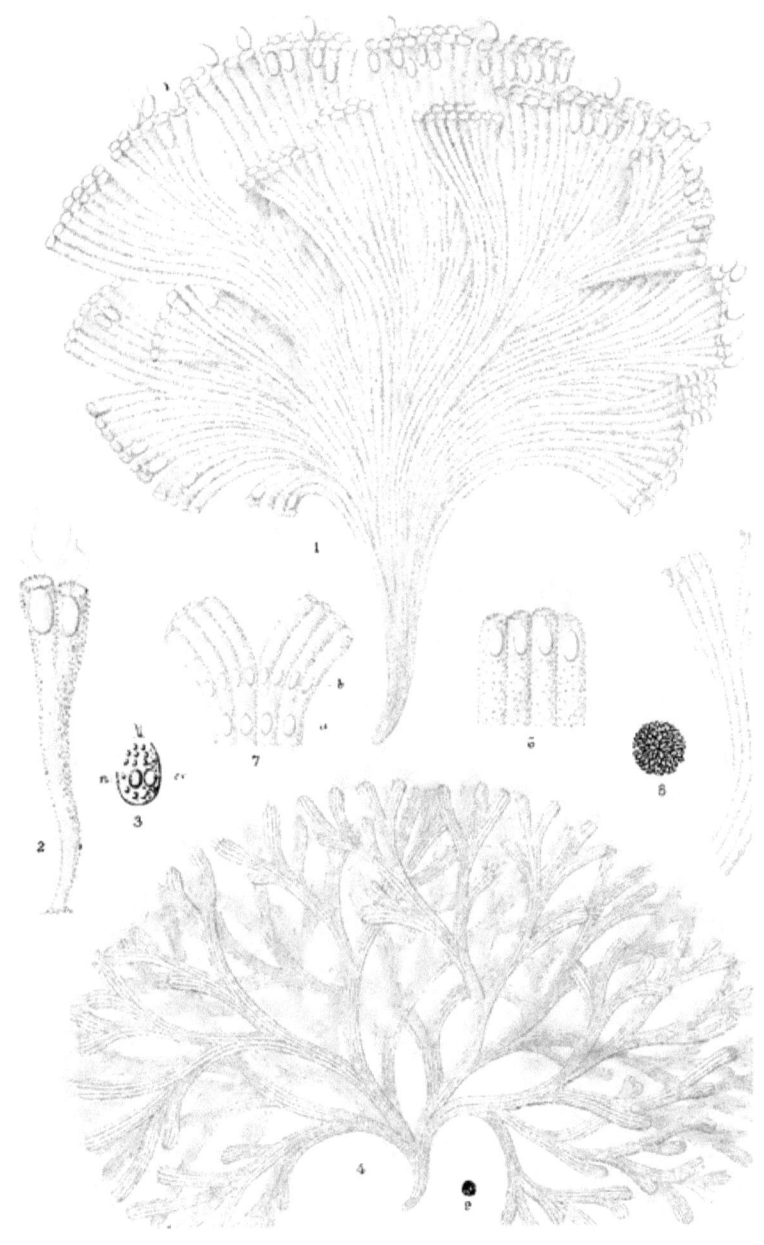

PLATE XVII.

EXPLANATION OF PLATE XVII.

FIG.

1–4. DENDROMONAS VIRGARIA, Weisse sp., vol. i. p. 266.—1, Adult colony-stock or zoodendrium, × 800; 2, younger and less ramified zoodendrium, as figured by Stein, 'Wiegmann's Archives,' 1849, as a probable young condition of *Epistylis anastatica*; 3, an isolated monad, × 2000; 4, colony of two monads in process of division by longitudinal fission.

5–7. CLADONEMA LAXA, S. K., vol. i. p. 265.—5, adult colony, × 1000; 6, a monad dividing by longitudinal fission; 7, a single monad ingesting a food-particle towards the posterior region of its periphery, × 3000.

8. DENDROMONAS PUSILLA, Schmard. sp., vol. i. p. 266, × 250.

9–11. ANTHOPHYSA SOCIALIS, From., vol. i. p. 272.—9 and 10, Colony-stocks, after De Fromentel, × 350; 11, head of colony-stock, including centrally an encysted monad, × 1500, after Bütschli.

12. CEPHALOTHAMNIUM CUNEATUM, S. K., vol. i. p. 273, × 1250.

13–26. ANTHOPHYSA VEGETANS, Müll. sp., vol. i. p. 267.—13–15, Branching colony-stocks and detached monad clusters, or cœnobia, as originally delineated by O. F. Müller, × 50; 16, a typical, erect, shortly branching colony-stock, with four terminal monad-clusters or cœnobia, × 400; 17, a portion of common stem, showing its compound or fibrous nature; 18 and 19, two colonies in which the monads, being fed with carmine, have after passing it through their bodies, incorporated the pigment from the points a, into the substance of the common stem, × 800; 20–22, sporocyst, showing progressive phases resulting in the liberation of the contents in the form of simple monadiform germs, × 800; 23, the empty sporocyst, showing its composition of an outer firmer and inner more delicate and elastic membrane; 24, monadiform germs liberated from sporocyst, × 2000; 25, subsequent resting condition of the same germs; 26, branching zoodendria derived from these resting spores, the monad clusters having become detached.

27–32. CEPHALOTHAMNIUM CÆSPITOSA, S. K., vol. i. p. 272.—27 and 28, Adult compound colony-stocks, × 1200; 29–31, single stalked monads, those in the last two figures emitting pseudopodia from within the distal region only, or from the entire surface of their periphery, × 1800; 32, a zooid dividing by transverse fission.—The size of the zooids in the descriptive account of this species has been accidentally set down as 1–5000″, instead of 1–3000.″

PLATE XVII

PLATE XVIII.

EXPLANATION OF PLATE XVIII.

FIG.

1–10. ANTHOPHYSA VEGETANS, Müll., vol. i. p. 267 (Stein).—1, Prolific colony-stock, with lax homogeneously-granulate branching stem, × 500; 2, detached monad cluster, or "cœnobium," seen from below; 3, cœnobium, showing its relationship to supporting pedicle, as seen in vertical optic section, × 1000; 4 and 5, free-swimming cœnobia, composed of a small number of monads only; 6, disintegrated cœnobium, the monads with tail-like posterior prolongations, × 1000; 7 and 8, independently motile monads, with pseudopodic processes, derived from breaking up of the adult cœnobium; 9, one such motile zooid, fixed by a posterior pseudopod-like extension; 10, early condition of sedentary colony-stock, bearing a single cœnobium of eight monads only.

11, 12. CLADOMONAS FRUTICULOSA, vol. i. p. 284 (Stein).—11, Adult tubular colony-stock or zoothecium, showing at a a detached monad, and at b two monads, the result of longitudinal fission, temporarily inhabiting the same tube, × 650; 12, terminal branchlet of variety having dark-coloured band-like internodes.

13–19. BICOSŒCA LACUSTRIS, vol. i. p. 275, J.-Clk.—13, Adult monad in lorica, × 1250; 14 and 15, successive phases of transverse fission; 16, distally separated, free-swimming integer of fissive process, presenting a Heteromita-like aspect; 17, monad with spirally-coiled flagella, retracted within its lorica; 18, lorica with sporular contents; 19, adult monad in lateral aspect, showing eccentric development of thread-like footstalk, and lip-like prolongation of the anterior region, × 1250.

20. BICOSŒCA CURVIPES, S. K.—Contained monad dividing by longitudinal fission or conjugating with an externally derived zooid, × 1500.

21, 22. BICOSŒCA GRACILLIPES, J.-Clk., vol. i. p. 276.—21, normal form; 22, example with shorter pedicle and more exsert zooid, × 1250.

23. BICOSŒCA TENUIS, S. K., vol. i. p. 276, × 1500.

24. HEDRÆOPHYSA BULLA, S. K., vol. i. p. 274, × 1500.

25–29. BICOSŒCA POCILLUM, S. K., vol. i. p. 277.—25, Typical adult form, × 1000; 26, monad dividing by transverse fission; 27 and 28, obtusely and acuminately pointed free-swimming monads derived from the fissive process; 29, a similarly derived monad adherent by its posterior extremity, but as yet wanting the characteristic protective lorica.

30, 31. DIPLOMITA SOCIALIS, S. K., vol. i. p. 289.—30, Social group attached to a confervoid filament; 31, two monads within their stalked loricæ, × 1500, at e eye-like pigment-spots.

32. STYLOBRYON EPISTYLOIDES, S. K., vol. i. p. 279, × 1500.

33–35. CEPHALOTHAMNIUM CÆSPITOSA, S. K., vol. i. p. 272 (Stein).—33, Adult colony-stock, × 1000; 34 and 35, free-swimming and attached isolated monads.

PLATE XIX.

EXPLANATION OF PLATE XIX.

FIG.
1-14. TREPOMONAS AGILIS, Duj., vol. i. p. 300.—1-8, Animalcule from various points of view, and with the body straight or twisted upon itself, as represented by Stein, at 7 an example dividing by longitudinal fission, × 650; 9, characteristic aspect as given by Perty; 10-12, other phases, as delineated by O. F. Bütschli, the arrows in the last instance indicating the course followed by the circulatory motion of the endoplasm; 13 and 14, young animalcules after Stein.

15. CHLORASTER AGILIS, S. K., vol. i. p. 317, × 1250.

16-19. CALLODICTYON TRICILIATUM, Carter, vol. i. p. 307.—17, an animalcule enveloping a vegetable filament, × 400 (Carter).

20. CHLORASTER TETRARHYNCHUS, Schm., vol. i. p. 316, × 350 (Stein).

21, 22. CHLORASTER GYRANS, Ehr., vol. i. p. 316, × 250 (Stein).

23-25. TRIMASTIX MARINA, S. K., vol. i. p. 312.—23 and 24, lateral, 25, oblique view, × 1250.

26, 27. TETRAMITUS SULCATUS, St., vol. i. p. 314, × 430 (Stein).

28, 29. TETRASELMIS CORDIFORMIS, Carter sp., vol. i. p. 315, × 450 (Stein).

30-32. TRICHOMONAS BATRACHORUM, Pty., vol. i. p. 308, × 650 (Stein).

33, 34. TRICHOMONAS VAGINALIS, Duj., vol. i. p. 309, × 1000 (Dujardin).

35-41. DALLINGERIA DRYSDALI, S. K., vol. i. p. 310 (Dallinger and Drysdale); 35, animalcule fixed by two posterior flagella, and showing at *a* and *b* respective positions occupied by the body during the alternate extensions and spiral contractions of these appendages, × 2000; 36, free-swimming animalcule, seen from above; 37, example dividing by longitudinal fission; 38, animalcule with posterior flagella retracted and about to enter upon an encysted state; 39, encystment; 40, sporocyst bursting and liberating clouds of microspores, × 2000; 41, triflagellate monads developed from the liberated microspores.

42-48. TETRAMITUS ROSTRATUS, Pty., vol. i. p. 313 (43, 47, and 48 after Stein, the remainder after Dallinger and Drysdale).—42, Normal free-swimming animalcule, × 1000; 43, ventral view, showing groove-like channel; 44 and 45, successive phases of longitudinal fission, accompanied by the assumption of an irregular amoebiform contour; 46, biflagellate animalcule derived by fission with flagella in process of further subdivision, so as to reproduce the normal number; 47 and 48, young animalcules, × 650.

49, 50. TETRAMITUS DESCISSUS, Pty., vol. i. p. 314, × 1500 (Bütschli).

51. LOPHOMONAS STRIATA, Bütschli, vol. i. p. 322, × 800 (Bütschli).

52-54. LOPHOMONAS BLATTARUM, St., vol. i. p. 321.—52 after Bütschli, 53 and 54, after Stein, × 650.

55. HEXAMITA ROSTRATA, St., vol i. p. 320, × 650.

56-59. HEXAMITA INFLATA, Duj., vol. i. p. 319 (56-58 after Stein).—56, normal adult animalcule, × 650; 57, animalcule about to divide by longitudinal fission; 58, example with posterior flagella entangled in floccose matter; 59, animalcule fixed by the extremities of its posterior flagellá, and gyrating upon the same, as observed by the author, × 1000.

60-62. HEXAMITA INTESTINALIS, Duj., vol. i. p. 318.—60, free-swimming animalcule, × 1000; 61, example anchored by the posterior flagella, and actively vibrating the four anterior ones; 62, an example with toothed antero-lateral borders and pseudopodic posterior prolongations (Stein).

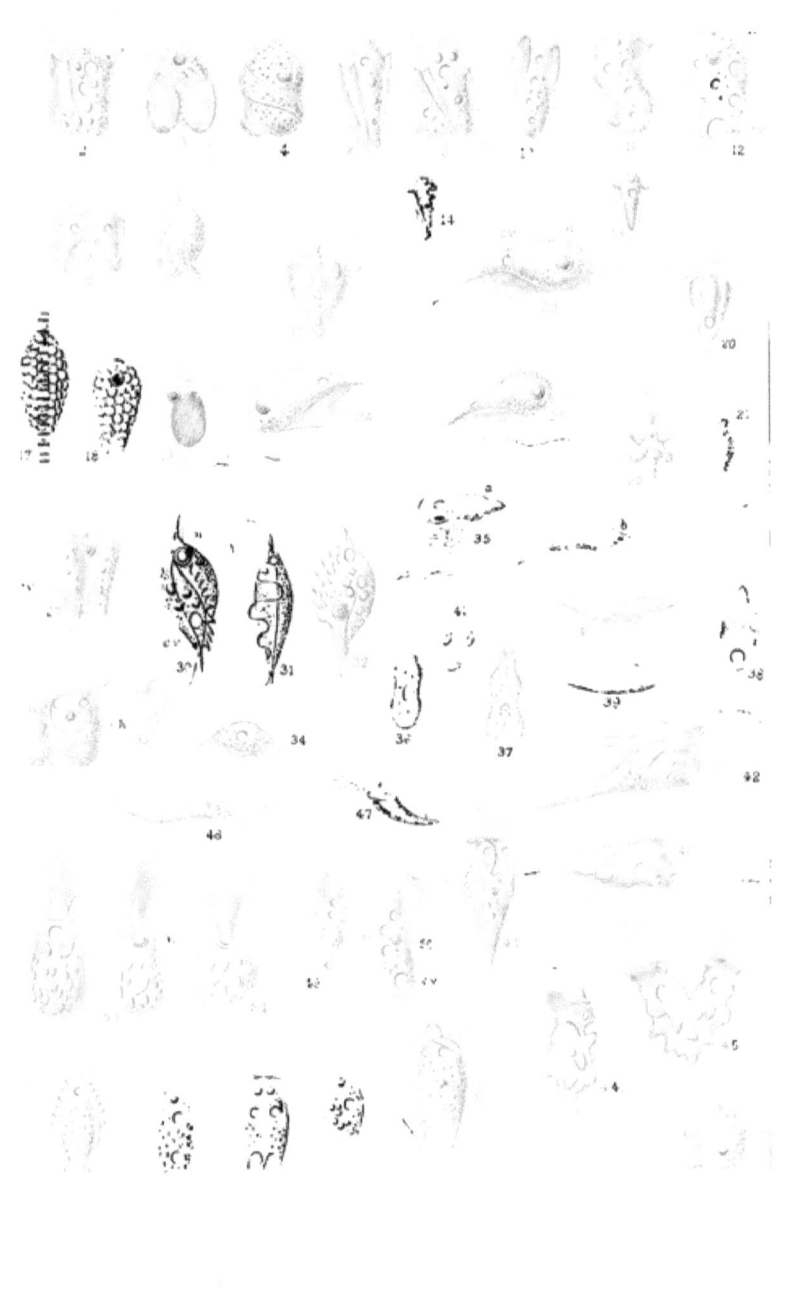

PLATE XX.

EXPLANATION.

FIG.
1. PARAMONAS GLOBOSA, From. sp., vol. i. p. 370, × 400 (Fromentel).
2. PARAMONAS STELLATA, From. sp., vol. i. p. 370, × 400 (Fromentel).
3. PETALOMONAS MEDIOCANELLATA, St., vol. i. p. 371, × 350 (Stein).
4. PETALOMONAS SINUATA, St., vol. i. p. 372, × 400 (Stein).
5, 6. PETALOMONAS ABSCISSA, Duj. sp., vol. i. p. 371, dorsal and ventral views, × 500 (Stein).
7. PETALOMONAS ERVILIA, St., vol. i. p. 372, × 400 (Stein).
8, 9. PETALOMONAS IRREGULARIS, S. K., vol. i. p. 372, × 1250.
10–12. ATRACTONEMA TERES, St., vol. i. p. 373.—11, Animalcule dividing by longitudinal fission; 12, example without flagellum, and with the endoplast enlarged and broken up into germinal elements, × 640.
13, 14. PHIALONEMA CYCLOSTOMUM, St., vol. i. p. 373, × 500.—14, Variety with spirally ribbed cuticle (Stein).
15, 16. MENOIDIUM PELLUCIDUM, Pty., vol. i. p. 374.—15, As observed by the author; 16, after Stein, × 600.
17–21. ASTASIA TRICHOPHORA, Ehr. sp., vol. i. p. 376.—17, Typical adult zooid, after Bütschli, × 600; 18, example with internal germ-like bodies, after Carter; 19–21, various polymorphic forms.
22. COLPODELLA PUGNAX, Cienk., vol. i. p. 378.—Three animalcules parasitically attached to a cell of *Protococcus*, × 500 (Cienkowski).
23. SPIROMONAS DISTORTA, Duj. sp, vol. i. p. 298, × 600 (Dujardin).
24–25. EUGLENA ACUS, Ehr., vol. i. p. 383.—25 and 26, after Stein, × 300.
26. EUGLENA OXYURIS, Schm., vol. i. p. 383, × 200 (Stein).
27, 28. EUGLENA SPIROGYRA, Ehr., vol. i. p. 382.—27, Normal extended animalcule, as observed by the author, at *a a*, amylaceous corpuscles, × 240; 28, contracted phase (Stein).
29–51. EUGLENA VIRIDIS, Ehr., vol. i. p. 381.—29, Animalcule with a bulbous distal termination to the flagella, after W. H. Robson; 30–33, polymorphic contours assumed at will by the same animalcule, × 200; 34, typical form, × 600; 35, anterior extremity, with *o f*, funicular oral fossa, *c v*, contractile vesicle, and *e*, eye-like pigment-spot; 36, animalcule with large ovate internally developed germs; 37, an example with enclosed irregularly shaped amœbiform germinal bodies; 38 and 39, transparent varieties (*E. hyalina*,

EXPLANATION OF PLATE XX. (*continued*).

FIG.

29–51 (*continued*). Ehr.) containing large central modified endoplasts or germ-spheres (after Stein); 40 and 41, similar animalcules giving exit by rupture of their germ-spheres to swarms of minute monadiform germs (Stein); 43, an animalcule emerging from a temporary encystment (Stein); 43, an example of fission as accomplished during the encysted state (Stein); 44–48, various phases of encystment accompanied with the production of internal germ-spheres, (Stein); 49 and 50, examples of encystment observed by the author, in which the entire endoplasmic contents have become divided into spheroidal spore-like bodies, at 50 a double sporocyst formed apparently by the conjugation of two animalcules; 51, repent amœbiform germs developed from the preceding sporocysts, × 600.

52, 53. EUGLENA DESES, Ehr., vol. i. p. 383, × 500.—At 53, a young non-flagellate zooid (Stein).

54–56. EUGLENA TUBA, Carter, vol. i. p. 385.—54, Normal free-swimming animalcule, × 300; 55, flask-shaped encystments associated in mucous reticulation; 56, a single flask-shaped body, more highly magnified (Carter).

57. EUGLENA VIRIDIS dividing by longitudinal fission, × 300 (Carter).

58. EUGLENA ZONALIS, Carter, vol. i. p. 388, × 300 (Carter).

59. CŒLOMONAS GRANDIS, Ehr. sp., vol. i. p. 393, × 400 (Stein).

60–62. RAPHIDOMONAS SEMEN, Ehr. sp., vol. i. p. 392.—At 61, example with hair-like trichocysts extended, × 400 (Stein).

63. AMBLYOPHIS VIRIDIS, Ehr., vol. i. p. 386, × 200 (Ehrenberg).

64. EUGLENA AGILIS, Carter, vol. i. p. 384.—Encysted animalcule, having by the abnormal process of serial subdivision become separated into four lineally disposed, equal-sized segment masses, × 300 (Carter).

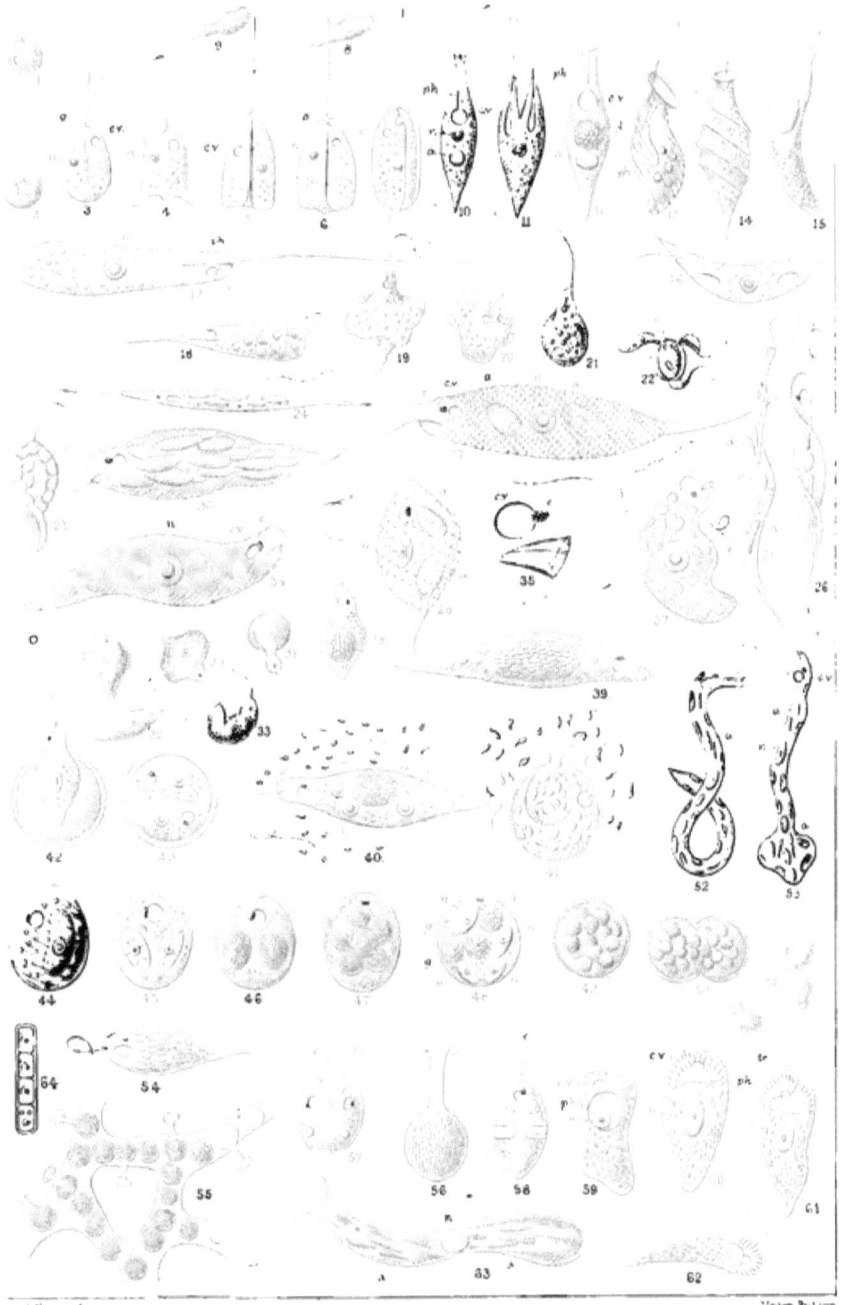

PLATE XXI.

EXPLANATION.

FIG.
1. PHACUS TRIQUETER, Ehr. sp., vol. i. p. 387.—*a a*, Amylaceous corpuscles; *e*, eye-like pigment-spot, × 600 (Stein).
2–5. PHACUS PLEURONECTES, Ehr. sp., vol. i. p. 386 (Stein).—2, Normal animalcule, × 600; 3, example with the nucleus or endoplast divided into four fragments, those at *g, g* being further separated into germinal elements, constituting the so-called "germ-sacs" or "germ-spheres" of Professor Stein ; 4, an instance of transverse fission ; 5, indurated membrane or carapace left after the decay of the enclosed body-substance.
6, 7. PHACUS LONGICAUDUS, Ehr. sp., vol. i. p. 387 (Stein).—6, More normal form; 7, spirally twisted non-flagellate example, × 300.
8, 9. CHLOROPELTIS HISPIDULA, Eichwald, vol. i. p. 388.—8, Lateral, 9, ventral aspect, × 600 (Stein).
10. PHACUS PYRUM, Ehr. sp., vol. i. p. 387, × 600 (Stein).
11–13. CHLOROPELTIS OVUM, Ehr., vol. i. p. 388.—*a a*, Amylaceous corpuscles ; 13, subcylindrical variety, × 600 (Stein).
14–16. TRACHELOMONAS VOLVOCINA, Ehr., vol. i. p. 389 (Stein).—14, Normal type, × 500 ; 15, animalcule with flagellum retracted, emerging from its lorica; 16, example bursting and liberating monadiform germs.
17. TRACHELOMONAS RUGULOSA, St., vol. i. p. 389, × 500 (Stein).
18, 19. TRACHELOMONAS LAGENELLA, Ehr. sp., vol. i. p. 389 (Stein).—19, Encysted example with body divided by transverse fission, × 500.
20. TRACHELOMONAS CYLINDRICA, Ehr., vol. i. p. 390, × 600 (Stein).
21–23. TRACHELOMONAS HISPIDA, Pty. sp., vol. i. p. 390 (Stein).—21, Necked variety, × 500; 22, neckless variety, with animalcule emerging from its lorica ; 23, example with lorica invested with an outer mucilaginous covering.
24. TRACHELOMONAS CAUDATA, Ehr. sp., vol. i. p. 391, × 500 (Stein).
25. TRACHELOMONAS ARMATA, Ehr. sp., vol. i. p. 390, × 500 (Stein).
26. TRACHELOMONAS ACUMINATA, Schmarda, vol. i. p. 391, × 450 (Stein).
27. TRACHELOMONAS EURYSTOMA, St., vol. i. p. 390, × 500 (Stein).
28, 29. ASCOGLENA VAGINICOLA, St., vol. i. p. 393 (Stein).—28, Normal form, × 500 ; 29, example divided by transverse fission, the anterior segment emerging from the parent lorica.

EXPLANATION OF PLATE XXI. (continued).

Fig.

30-32. COLACIUM CALVUM, St., vol. i. p. 395 (Stein).—30, Social colony-stock, × 400; 31 and 32, detached, free-swimming animalcules in extended and contracted states.

33. COLACIUM ARBUSCULA, St., vol. i. p. 394.—Tree-like colony-stock, × 500 (Stein).

34-38. COLACIUM VESICULOSUM, Ehr., vol. i. p. 395 (Stein).—34, Sedentary colony-stock, × 600; 35, detached, free-swimming animalcule; 36-38, successive developmental phases following on the attachment of a motile zooid, and resulting, through further subdivision, in the building up of a more or less extensive sedentary stock.

39-41. COLACIUM STEINII, S. K., vol. i. p. 395.—39, Erect colony-stock, × 350; 40, arborescent stock, × 700, showing variety of contracted or contorted forms assumed at will by the component animalcules; 41, animalcule in an encysted state, with the body-contents divided into sporular elements.

42. CRYPTOGLENA CONICA, Ehr., vol. i. p. 419, × 500 (Ehrenberg).

43-45. CRYPTOGLENA ANGULOSA, Carter, vol. i. p. 420 (Carter).—43, Normal aspect; 44, example enclosing four endogenously developed germinal bodies; 45, lateral aspect, × 500.

46-51. DISTIGMA PROTEUS, Ehr., vol. i. p. 418 (Stein).—46, Normal biflagellate animalcule, × 500; 47 and 48, old, non-flagellate, and repent zooids (*Proteus tenax*, O. F. Müller?); 49 and 50, young free-swimming animalcules; 51, still younger monoflagellate example (*Monas punctum*, Ehr.).

52, 53. ZYGOSELMIS NEBULOSA, Duj., vol. i. p. 417 (Stein), × 1000.—53, Example with body distended by ingested diatoms.

54-59. EUTREPTIA VIRIDIS, Pty., vol. i. p. 416.—54 and 55, Normal free-swimming animalcules, × 250; 56, repent example; 57, sporocyst burst and liberating green amœbiform germs, × 400; 58, amœbiform germs more highly magnified; 59, young monoflagellate zooids, × 250.

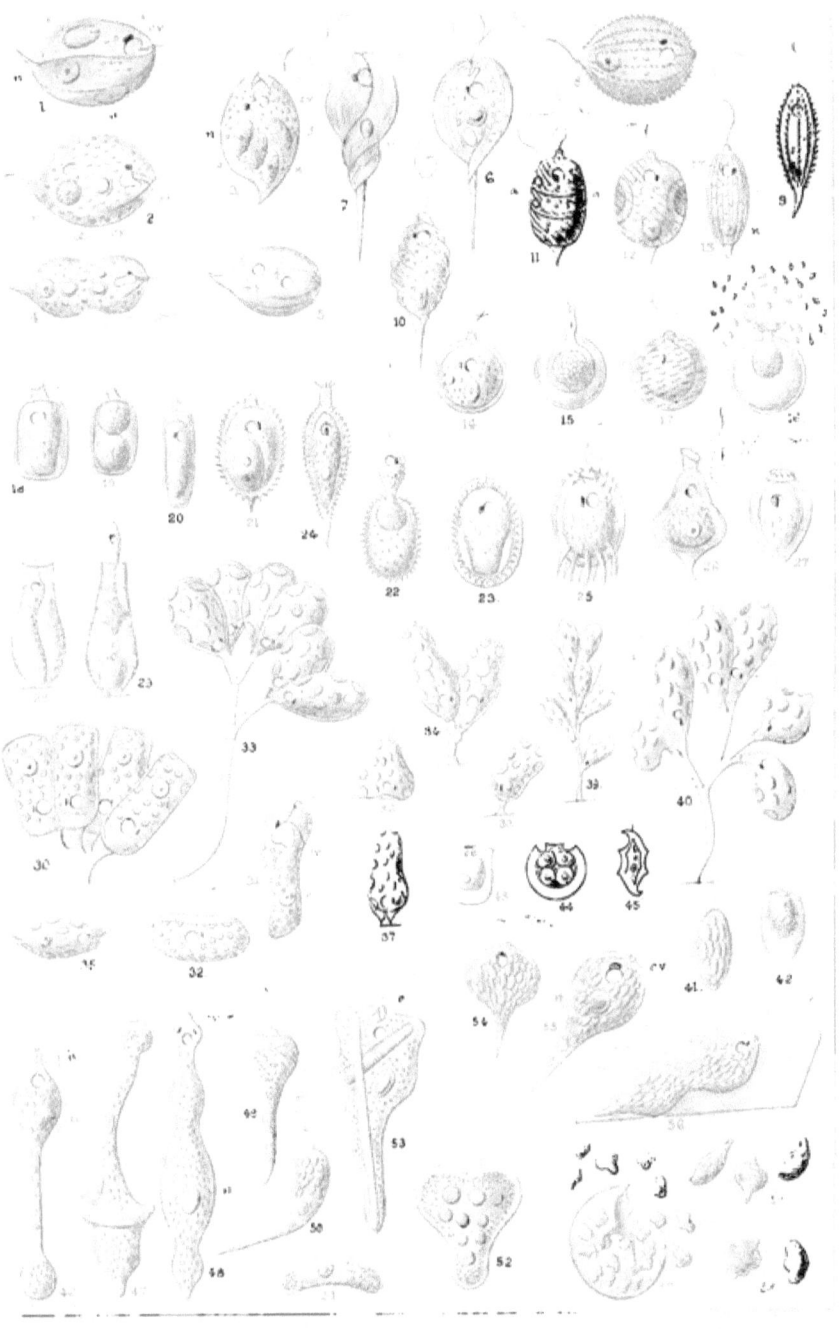

PLATE XXII.

EXPLANATION OF PLATE XXII.

FIG.

1, 2. CHLOROMONAS PIGRA, Ehr. sp., vol. i. p. 401.—Vertical and profile aspects, × 1500 (Stein).

3-7. CHRYSOMONAS OCHRACEA, Ehr. sp., vol. i. p. 402 (Stein).—3 and 4, Normal free-swimming animalcules, × 600; 5-7, resting or encysted conditions, accompanied by multiplication by segmentation into two, four, or eight spore-like daughter-cells.

8, 9. CHRYSOMONAS FLAVICANS, Ehr. sp., vol. i. p. 402 (Stein).—8, Normal free-swimming animalcule, × 600; 9, resting or encysted reproductive phase.

10. MICROGLENA PUNCTIFERA, Ehr., vol. i. p. 403, × 600 (Stein).

11-13. NEPHROSELMIS OLIVACEA, St., vol. i. p. 405, × 600 (Stein).

14-15. HYMENOMONAS ROSEOLA, St., vol. i. p. 408, × 600 (Stein).

16-18. CRYPTOMONAS OVATA, Ehr., vol. i. p. 404 (Stein).—16 and 17, Adult animalcules, × 600; 18, young example.

19-21. CRYPTOMONAS EROSA, Ehr., vol. i. p. 404 (Stein).—19, Adult animalcule enclosing large ovate germ-mass, × 600; 20, encysted animalcules; 21, example dividing by longitudinal fission.

22. STYLOCHRYSALIS PARASITA, St., vol. i. p. 405.—Social group attached to the cell-wall of a *Eudorina*, one example at *a* dividing by transverse fission, × 500 (Stein).

23. DINOBRYON JUNIPERINUM, Eichw., vol. i. p. 411, dimensions unrecorded (Eichwald).

24-26. UVELLA VIRESCENS, Ehr., vol. i. p. 406 (Butschli).—24, Socially united cluster of five zooids, × 1200; 25, example treated with acetic acid, showing central spheroidal nucleus or endoplast, and two laterally developed pigment-bands; 26, an animalcule dividing by longitudinal fission.

27. SYNURA (RHOIPESSA) GRIMSELINA, Pty. sp., vol. i. p. 412, × 300 (Perty).

28, 29. CHRYSOPYXIS BIPES, St., vol. i. p. 408 (Stein).—28, Colony fixed to a node of *Mougeotia*, × 600; 29, isolated example showing the two spine-like posterior prolongations of the lorica, by which adhesion to the confervoid alga is secured.

30-33. EPIPYXIS UTRICULUS, Ehr., vol. i. p. 409 (Stein).—30, Social colony, × 600, at *a* lorica with network-like markings, probably indicating the presence or previous existence of sporular elements; 31, a more abnormal and closely aggregated colony; 32, an isolated animalcule with its protective lorica more highly magnified; 33, an example dividing by oblique fission, as observed by the author.

34-40. DINOBRYON SERTULARIA, Ehr., vol. i. p. 409.—34, Compound detached polythecium, × 250; 35, a similar but smaller colonial aggregation, × 1000, at *a* an animalcule derived by recent fission, attached to the margin of the parent lorica, and constructing for itself by exudation a new protective sheath; 36, an isolated animalcule with its investing lorica, × 2500; *e*, coloured eye-like pigment-spot; *am*, amylaceous corpuscle; *st*, thread-like contractile pedicle or footstalk; *n*, nucleus or endoplast; *cv*, contractile vesicle; 37, example dividing by transverse fission (Bütschli); 38, encysted example (Bütschli); 39 and 40, examples of encystment, after Stein.

41. DINOBRYON STIPITATUM, St., vol. i. p. 410, × 500 (Stein).

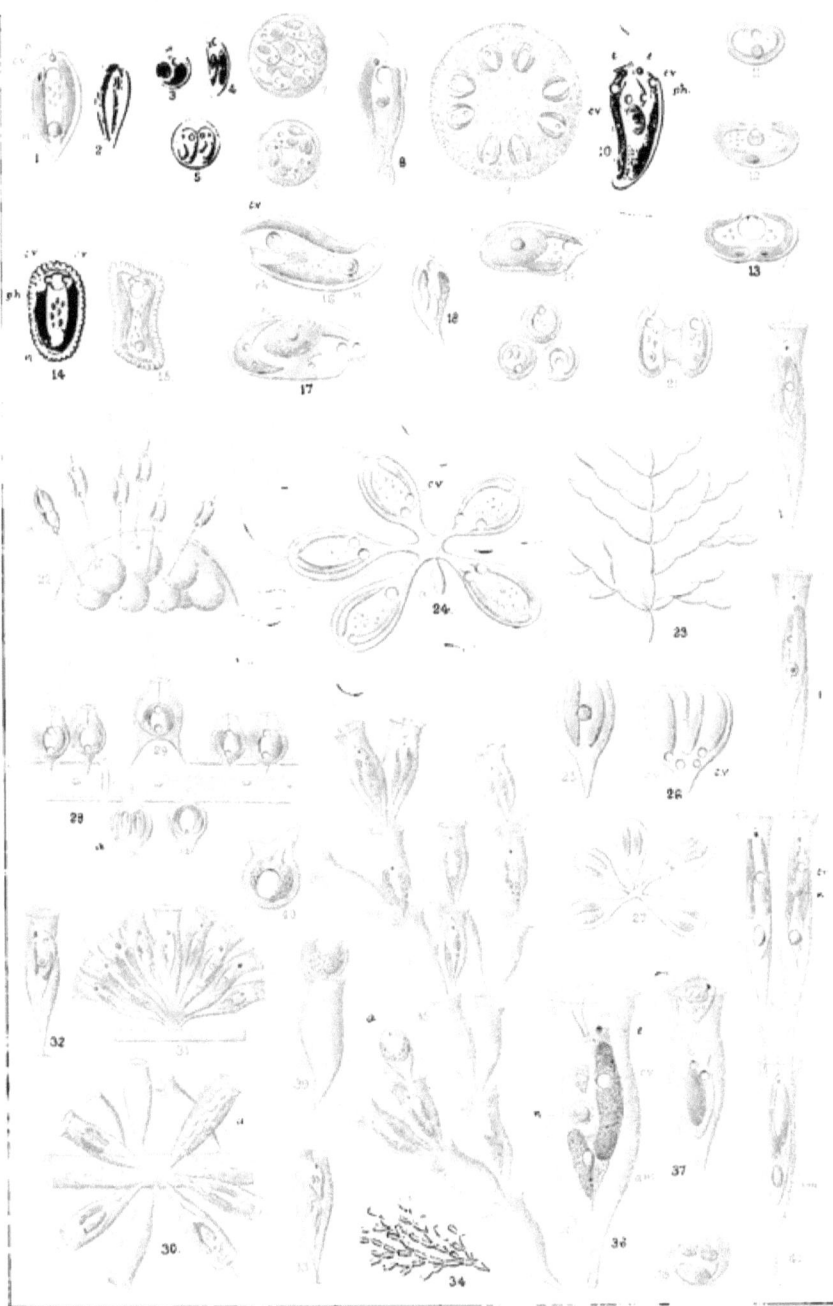

PLATE XXIII.

EXPLANATION OF PLATE XXIII.

FIG.

1–2. SYNURA UVELLA, Ehr., vol. i. p. 412 (Stein).—1, Adult spheroidal colony-stock, composed of animalcules possessing no eye-like pigment-specks, × 350; 2, smaller colony of variety having two anterior eye-like pigment-specks.

3. SYNCRYPTA VOLVOX, Ehr., vol. i. p. 413, × 200 (Stein).

4–15. UROGLENA VOLVOX, Ehr., vol. i. p. 414.—4, Subspheroidal colony-stock, × 150; 5, segment of a similar colony, more highly magnified, containing sporular elements in various phases of development; 6, a single animalcule; e, eye-like pigment-speck; cv, contractile vesicle; am, refringent amylaceous corpuscle; 7 and 8, isolated spores, × 1200; encysted zooid, with body-mass divided by segmentation into four sporular elements; 10 and 11, sporocysts fractured by artificial pressure, and liberating their contained sporular elements, × 400; 12, animalcule as represented by Stein; 13, three zooids of the same species, after Bütschli; 14–15, examples treated with acetic acid and carmine, after Bütschli and Stein.

16. SYNURA (UVELLA) FIMBRIATA, From. sp., × 400 (De Fromentel).

17–30. STYLOBRYON PETIOLATUM, Duj. sp., vol. i. p. 278.—17, Colony-stock of six zooids, as observed by the author, having very long, slender pedicles to their loricæ, × 400; 18, colony-stock, in which the pedicles of the distal series are rudimentarily developed; 19 and 20, animalcules in dorsal and profile view, as examined in living state, × 800; 21–23, example from another colony preserved with osmic acid, the longer flagellum being variously convoluted; 24 and 25, examples with entire body-substance separated into sporular elements; 26, isolated spores, × 1200; 27 and 28, colony-stocks of the same type, after Stein; 29 and 30, colony-stock and isolated animalcule of the same species, as figured and described by De Fromentel under the title of *Stylobryon insignis*.

PLATE XXIII

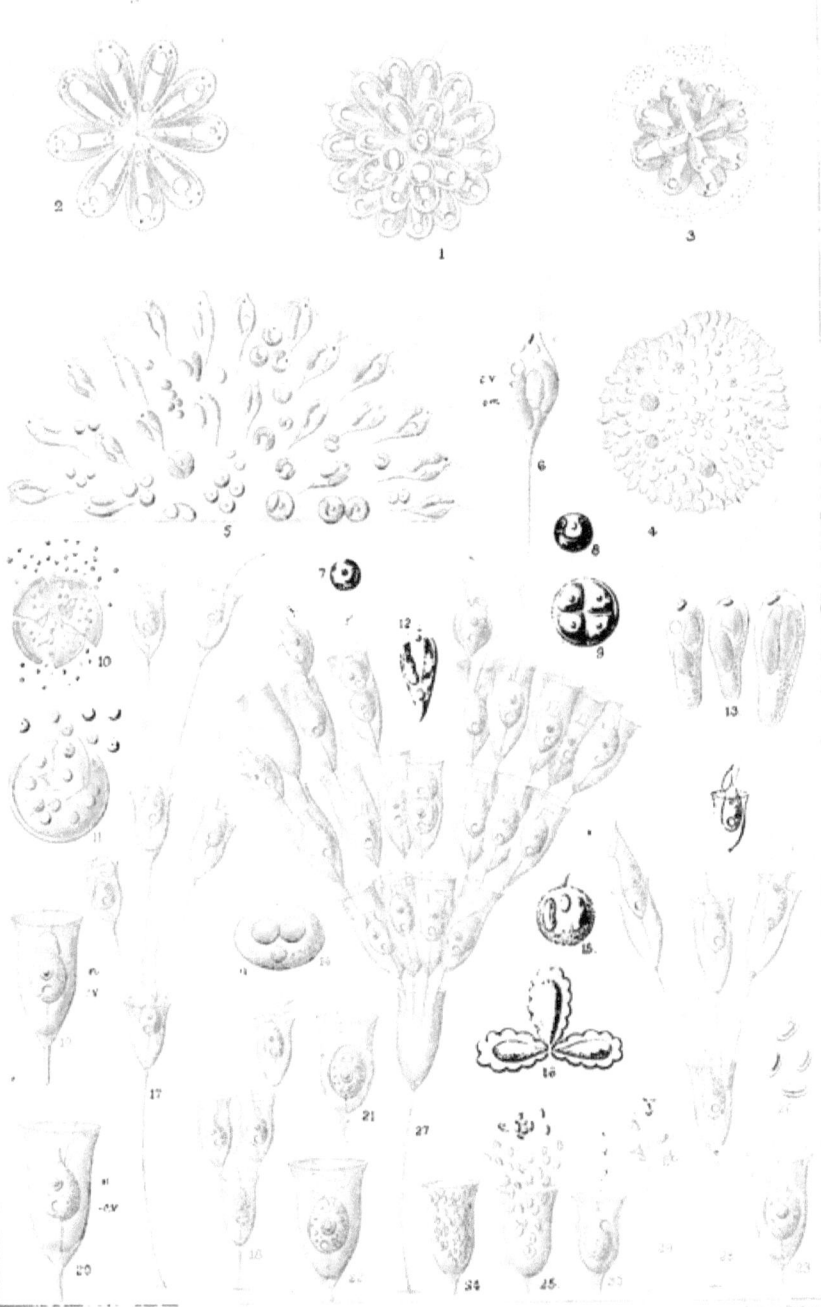

PLATE XXIV.

EXPLANATION.

FIG.
1-10. DIPLOMASTIX CAUDATA, Duj. sp., vol. i. p. 432.—1, Typical adult form, × 800 ; 2 and 3, examples with prolonged posterior extremities ; 4, conjugation of two animalcules ; 5, spore-groups, imbedded in decaying hay-fibre, at *a a* two uniflagellate, monadiform germs, produced from the same, × 1000 ; 6, sporocyst with four rounded macrospores, × 1000 ; 7, group of elongate spores, with cyst-wall dissolved, × 1000 : 8, sporocyst with germ emerging, after Stein ; 9 and 10, animalcule ingesting and having ingested a bacillus filament, after Stein.

11, 12. DIPLOMASTIX SALTANS, Ehr. sp., vol. i. p. 433, × 800 (Stein).

13. DIPLOMASTIX AFFINIS, S. K., vol. i. p. 433, × 800.

14, 15. HETERONEMA ACUS, Ehr. sp., vol. i. p. 430, × 400 (Stein).

16, 17. HETERONEMA GLOBULIFERUM, Ehr. sp., vol. i. p. 430, × 400 (Stein).

18-20. SPHENOMONAS OCTOCOSTATA, St. sp., vol. i. p. 439 (Stein).—18 and 19, Lateral and ventral aspects, × 400 ; 20, quiescent or encysted phase.

21-23. SPHENOMONAS QUADRANGULARIS, St., vol. i. p. 438 (Stein).—21, Normal adult type, × 400 ; 22, the same viewed vertically to its long axis ; 23, zooid dividing by longitudinal fission.

24, 25. ANISONEMA TRUNCATUM, St., vol. i. p. 435 (Stein).—24, Dorsal, 25, ventral, aspects, × 450.

26-30. ANISONEMA GRANDE, Ehr. sp., vol. i. p. 434.—26, Axial, and 27, dorsal, aspects, × 400, the four flagella present in the last figure indicate approaching fission (after Jas.-Clark) ; 28 and 29, dorsal and ventral aspects, the zooid in the first instance containing two germinal masses, × 450, after Stein ; 30, zooid after O. Bütschli.

31-34. ENTOSIPHON SULCATUM, Duj. sp., vol. i. p. 437 (Stein), × 400.—31, Zooid enclosing ovate germ-mass ; 32, living, and 34, dead examples, with their horny pharyngeal tubes protruding.

35, 36. ANISONEMA LUDIBUNDUM, S. K., vol. i. p. 435, × 1200.—35, Animalcule in lateral view discharging fæcal matter from its posterior extremity ; 36, dorsal aspect, showing the two anteriorly located contractile vesicles.

37-39. ANISONEMA INTERMEDIUM, S. K., vol. i. p. 436.—37 and 39, dorsal and lateral aspects, × 1200 ; 28, two young animalcules, possessing as yet only a single posterior anchoring flagellum, or gubernaculum, × 1200.

40-42. STERROMONAS FORMICINA, S. K., vol. i. p. 420.—40 and 41, Dorsal and lateral aspects, × 1200 ; 42, encysted state.

EXPLANATION OF PLATE XXIV. (continued).

FIG.

43–45. DINOMONAS TUBERCULATA, S. K. vol. i. p. 422.—43, Animalcule with prolonged caudal extremity, × 1800; 44, an animalcule devouring a smaller monad; 45, an example with ingested *Bacillus*.

46–48. DINOMONAS VORAX, S. K., vol. i. p. 422.—46, Normal adult type, × 1200; 47, more attenuate early condition; 48, example in the act of devouring a smaller monad.

49. CHILOMONAS AMYGDALUM, S. K., vol. i. p. 426, × 1800.

50. CHILOMONAS CYLINDRICA, Ehr. sp., vol. i. p. 425, × 500 (Bütschli).

51, 52. CHILOMONAS PARAMECIUM, Ehr., vol. i. p. 424 (Bütschli).—51, Normal adult form, × 650; 52, example dividing by longitudinal fission.

53–61. OXYRRHIS MARINA, Duj., vol. i. p. 427.—53, A reproduction of Dujardin's original figure; 54–56, right and left side aspects, as observed by the author, × 800; 57, an example of transverse fission; 58 and 59, empty membranous carapaces; 60 and 61, delineations of the same animalcule as given by Cohn.

62–64. ASTHMATOS CILIARIS, Sals., vol. i. p. 467, × 400 (Salisbury).

65, 66. TRICHONEMA HIRSUTA, From., vol. i. p. 469, × 400 (Fromentel).

67, 68. MITOPHORA DUBIA, Pty., vol. i. p. 469, × 350 (Perty).

69. STEPHANOMONAS LOCELLUS, From. sp., vol. i. p. 466, × 400 (Fromentel).

70, 71. HETEROMASTIX PROTEIFORMIS, J.-Clk., vol. i. p. 463.—70, extended, and 71, contracted, phases, × 500 (Jas.-Clark).

72, 73. MALLOMONAS PLOSSLII, Pty., vol. i. p. 464, × 800.

74. MALLOMONAS FRESENII, S. K., vol. i. p. 465, × 350.

PLATE XXV.

EXPLANATION OF PLATE XXV.

FIG.
1–5. PERIDINIUM TABULATUM, Ehr., vol. i. p. 448.—1, Variety with rounded poles or apices and no eye-like pigment-spot, ventral aspect, × 350 (Claparède and Lachmann); 2, variety with pointed apices, dorsal aspect (Ehr.); 3, separate reticulated plate from cuirass, × 600; 4, encysted example, with eye-like pigment-spot (C. & L.); 5, example with four pigment-spots, its cuirass cast off (C. & L.).
6, 7. PERIDINIUM APICULATUM, Ehr. sp., vol. i. p. 449, × 400 (Ehr.).
8–13. CERATIUM DIVERGENS, C. & L., vol. i. p. 453.—8 and 9, Empty cuirass, dorsal and ventral aspects, showing form and disposition of its component plates, × 500 (C. & L.); 10, encysted animalcule (C. & L.); 11, example with carapace cast off (C. & L.); 12 and 13, the same species as figured by Bailey under the title of *Peridinium depressum*.
14. PERIDINIUM ÆQUALIS, S. K., vol. i. p. 451.—Dimensions unrecorded (Will.-Suhm).
15, 16. PERIDINIUM SPINIFERUM, C. & L., vol. i. p. 449, × 400 (C. & L.).
17, 18. GYMNODINIUM FUSCUM, Ehr., vol. i. p. 443, × 300 (Ehr.).—18, Example of conjugation or longitudinal fission.
19, 20. GYMNODINIUM PULVISCULUS, Ehr. sp., vol. i. p. 443, × 500 (Ehr.).
21, 22. GLENODINIUM ACUMINATUM, Ehr., vol. i. p. 446, × 250 (Ehr.).
23. CERATIUM MICHAELIS, Ehr. sp., vol. i. p. 453, × 300 (Ehr.).
24. CERATIUM TRIPOS, Müll. sp., var. MACROCEROS, vol. i. p. 454, × 250 (C. & L.).
25. CERATIUM KUMAONENSE, Carter, vol. i. p. 458, × 190 (Carter).
26. CERATIUM LONGICORNE, Perty, vol. i. p. 457, × 300 (Carter).
27, 28. GLENODINIUM CINCTUM, Ehr. sp., vol. i. p. 446, × 400 (Ehr.).—28, An example of longitudinal fission or conjugation.
29, 30. PERIDINIUM sp., encysted examples, vol. i. p. 447, × 250 (C. & L.).
31, 32. CERATIUM FURCA, Ehr. sp., vol. i. p. 445, × 350.—32, Central region, showing more minute structure of the cuirass, × 600 (C. & L.).
33. CERATIUM TRIPOS, Müll. sp., normal short-armed type, × 250 (C. & L.).
34, 35. MELODINIUM UBERRIMUM, Allman sp., vol. i. p. 415.—34, Normal animalcule, × 300; 35, example dividing by transverse fission (Allman).
36. CERATIUM TRIPOS, Müll. sp., var. ARCTICUM, vol. i. p. 454, × 300.
37–39. PROROCENTRUM MICANS, Ehr., vol. i. p. 461.—37 and 38, Lateral and dorsal aspects; 39, empty cuirass, × 303 (C. & L.).
40. CERATIUM FUSUS, Ehr. sp., vol. i. p. 456, × 300 (C. & L.).
41. PERIDINIUM RETICULATUM, C. & L., vol. i. p. 449, × 300 (C. & L.).
42. DINOPHYSIS VENTRICOSA, C. & L., vol. i. p. 459, × 350 (C. & L.).
43. DINOPHYSIS ACUMINATA, C. & L., vol. i. p. 459, × 300 (C. & L.).
44. DINOPHYSIS OVATA, C. & L., vol. i. p. 460, ventral aspect, × 350 (C. & L.).
45, 46. AMPHIDINIUM OPERCULATUM, C. & L., vol. i. p. 461, × 300 (C. & L.).
47–50. PERIDINIA sp., encystments, vol. i. p. 448, with contents variously divided, that at 49 with the contained protoplasmic mass separated into eight minute naked Peridinia, × 300 (after Claparède and Lachmann).
51, 52. DIMASTIGOAULAX CORNUTUM, Ehr. sp., vol. i. p. 462, front and lateral aspects, × 300.
53. GYMNODINIUM ROSEOLUM, Schmarda sp., vol. i. p. 444, × 350 (Schmarda).
54. GYMNODINIUM INERME, Schmarda sp., vol. i. p. 444, × 600 (Schmarda).
55–57. PERIDINIUM TABULATUM, Ehr., vol. i. p. 448, brown variety with cleft anterior border, × 300.—55, Ventral, 56, dorsal, 57, lateral aspects, the example at 56 enclosing a recurved band-like endoplast.
58, 59. GYMNODINIUM LACHMANNI, S. K., vol. i. p. 444.—At 58 an example dividing by longitudinal fission, × 300 (C. & L.).
60, 61. GYMNODINIUM MARINUM, S. K., vol. i. p. 444, × 600.—At 61 an example devouring a smaller monad.
62. CERATIUM BICORNE, Schmarda sp., vol. i. p. 453, × 400 (Schmarda).

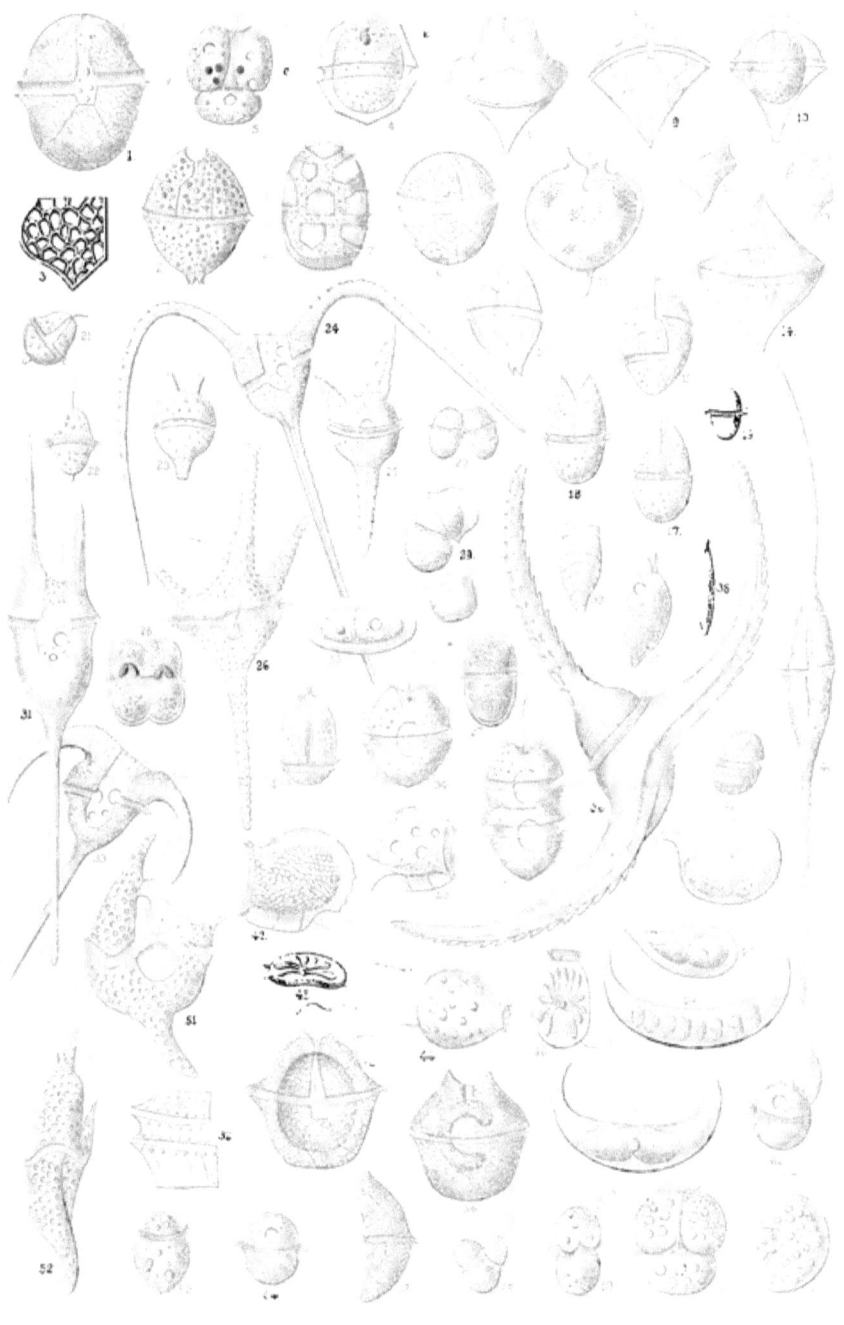

PLATE XXVI.

EXPLANATION.

FIG.
1–9. OPALINA RANARUM, Purkinge, vol. ii. p. 559.—1, Adult animalcule, × 100; 2–5, successive phases of segmentation terminating in the production of minute ovate zooids possessing but a few nuclei or endoplasts, × 100; 6, succeeding encysted condition of No. 5, × 200: 7 and 8, successive developmental phases of zooid that has re-emerged from the encysted state, the one at 7 possessing but a single endoplast, and that at 8 three such structures, the lowermost of which, at a, is in the act of subdividing; 9, a young zooid with ragged pseudopodium-like lateral extensions (1–8 after Ernst Zeller, 9 after T. W. Engelmann).
10–11. OPALINA OBTRIGONA, Stein, vol. ii. p. 562.—10, Adult zooid, × 100; 11, minute encystment, × 200 (Ernst Zeller).
12. ANOPLOPHRYA NAIDOS, Duj. sp., vol. ii. p. 563.—Adult zooid, × 200 (Ray Lankester).
13. ANOPLOPHRYA FILUM, C. & L., vol. ii. p. 567, × 120 (Claparède & Lachmann).
14. ANOPLOPHRYA PROLIFERA, C. & L., vol. ii. p. 564.—Adult example with five incompletely separated posterior segments, × 120 (Clap. & Lach.).
15. HOPLITOPHRYA LUMBRICI, Duj. sp., vol. ii. p. 571.—Adult zooid in the act of dividing by transverse fission, at u, u, horny uncini, × 200 (Stein).
16–18. OPALINA DIMIDIATA, Stein, vol. ii. p. 561.—16, Adult animalcule, × 100; 17, minute zooid, produced through successive fission of adult unit, commencing to subdivide again in a longitudinal direction; 18, elongate zooid, containing but a single endoplast, recently emerged from a minute encystment (Ernst Zeller).
19. OPALINA INTESTINALIS, Ehr. sp., vol. ii. p. 562, × 100 (Ernst Zeller).
20. OPALINA RANARUM, Purk., vol. ii. p. 559.—Fragment of cuticular fibrillæ treated with acetic acid, × 200 (Zeller).
21. ANOPLOPHRYA CLAVATA, Leidy sp., vol. ii. p. 566, × 150 (Leidy).
22. HOPLITOPHRYA RECURVA, C. & L. sp., vol. ii. p. 573, × 130 (Clap. & Lach.).
23, 24. OPALINA CAUDATA, Zeller, vol. ii. p. 563.—Dorsal and lateral aspects, × 120 (Zeller).
25. ANOPLOPHRYA COCHLEARIFORMIS, Leidy sp., vol. ii. p. 566, × 150 (Leidy).
26, 27. ANOPLOPHRYA MYTILI, Quenn. sp., vol. ii. p. 565.—Lateral and dorsal aspects, × 250 (Quennerstedt).
28–30. PARAMÆCIUM AURELIA, Müller, vol. ii. p. 483.—28 and 29, Lateral and ventral aspects, × 200; 30, two conjugating zooids.

EXPLANATION OF PLATE XXVI. (*continued*).

FIG.
31-32. PARAMÆCIUM BURSARIA, Ehr. sp., vol. ii. p. 486.—Dorsal and ventral aspects × 250. The arrows in each case indicate the direction of the endoplasmic current or cyclosis.
33. CONCHOPHTHIRUS ANODONTÆ, Ehr. sp., vol. ii. p. 490, × 200 (Engelmann).
34, 35. CONCHOPHTHIRUS STEENSTRUPII, Stein, vol. ii. p. 490.—Ventral and lateral aspects, × 200 (Quennerstedt).
36. PRORODON NIVEUS, Ehr., vol. ii. p. 492, × 75 (Ehrenberg).
37. CYRTOSTOMUM LEUCAS, Ehr. sp., vol. ii. p. 497, × 150 (Ehrenberg).
38. ISOTRICHA MICROSTOMUM, C. & L. sp., vol. ii. p. 498, × 250 (Clap. & Lach.).
39, 40. PLACUS STRIATUS, Cohn, vol. ii. p. 490.—Ventral and lateral aspects, × 450 (Cohn).
41. NASSULA AMBIGUA, Stein, vol. ii. p. 495, × 200 (Stein).
42. NASSULA ORNATA, Ehr., vol. ii. p. 494, × 100 (Ehrenberg).
43. PRORODON EDENTATUS, C. & L., vol. ii. p. 493, × 260 (Clap. & Lach.).
44. PRORODON MARGARITIFER, C. & L., vol. ii. p. 493, × 120 (Clap. & Lach.).
45. HOLOPHRYA OVUM, Ehr., vol. ii. p. 498, × 200 (Clap. & Lach.).
46. HOLOPHRYA LATERALIS, S. K., vol. ii. p 500, × 100 (Carter).
47. LOXOCEPHALUS GRANULOSUS, S. K., vol. ii. p. 489, × 300.
48. CHASMATOSTOMA RENIFORME, Eng., vol. ii. p. 540, × 200 (Engelmann).
49. PRORODON NIVEUS, Ehr., vol. ii. p. 492.—Pharyngeal rod-fascicle, × 120 (Ehrenberg).
50. NASSULA ORNATA, vol. ii. p. 494.—Pharyngeal rod-fascicle, × 200 (Ehrenberg).
51-53. ENCHELYODON FARCTUS, C. & L., vol. ii. p. 503.—51, Extended, 52, contracted states, × 120; 53, contractile vesicle at full diastole, with its lateral sinuses and at *a* central pore-like opening (Wrzesniowski).
54. HELICOSTOMA OBLONGA, Cohn., vol. ii. p. 501, × 150 (Cohn).
55-58. OTOSTOMA CARTERI, S. K., vol. ii. p. 500.—55 and 56, Adult animalcules, that at 56 showing the band-like endoplast and stellate contractile vesicles, × 100; 57 and 58, successive phases of subdivision into sporular elements (Carter).
59, 60. HOLOPHRYA TARDA, Quenn., vol. ii. p. 499.—Extended and contracted states, × 200 (Quennerstedt).
61, 62. TRACHELOPHYLLUM APICULATUM, Perty sp., vol. ii. p. 502.—Extended and contracted states, × 200 (Wrzesniowski).
63, 64. OPHRYOGLENA ATRA, Ehr., vol. ii. p. 532.—Lateral and ventral views, × 200 (Ehrenberg).
65, 66. PANOPHRYS FLAVICANS, Ehr. sp., vol. ii. p. 534.—Ventral and lateral views, × 150 (Ehrenberg).
67, 68. CYCLOTRICHA CITREUM, C. & L. sp., vol. ii. p. 535.—Lateral and ventral aspects, × 230 (Clap. & Lach.).
69. PLAGIOPYLA (?) CARTERI, S. K., vol. ii. p. 538, × 100.
70. PLAGIOPYLA (?) FUSCA, Quenn. sp., vol. ii. p. 539, × 150 (Quennerstedt).

PLATE XXVI

PLATE XXVII.

EXPLANATION.

FIG.
1. UROTRICHA LAGENULA, Ehr. sp., vol. ii. p. 505, × 300.
2. UROTRICHA FARCTA, C. & L., vol. ii. p. 505, × 350 (Claparède & Lachmann).
3, 4. COLEPS HIRTUS, Ehr., vol. ii. p. 506.—3, Adult animalcule, × 400; 4, example dividing by transverse fission.
5. COLEPS FUSUS, C. & L., vol. ii. p. 507, × 350 (Clap. & Lach.).
6. COLEPS UNCINATUS, C. & L., vol. ii. p. 507, × 350 (Clap. & Lach.).
7. PLAGIOPOGON COLEPS, Ehr. sp., vol. ii. p. 508, × 250 (Perty).
8-10. POLYKRIKOS SCHWARTZII, Bütschli, vol. ii. p. 508.—8, Adult example, × 300; 9 and 10, supposed trichocysts with thread-like filament in the retracted and extended states, × 600 (Bütschli).
11-13. METACYSTIS TRUNCATA, Cohn, vol. ii. p. 511.—11 and 12.—Adult examples, × 400; young zooids as yet devoid of transverse annulation (Cohn).
14. ENCHELYS ARCUATA, C. & L., vol. ii. p. 510, × 300 (Clap. & Lach.).
15. ENCHELYS FARCIMEN, Ehr., vol. ii. p. 510, × 600.
16, 17. ANOPHRYS SARCOPHAGA, Cohn, vol. ii. p. 512, × 400 (Cohn).
18. PERISPIRA OVUM (?) Stein, vol. ii. p. 511.—Dimensions unrecorded (Carter).
19-23. COLPODA CUCULLUS, Ehr., vol. ii. p. 512.—19, Adult animalcule, × 200; 20-23, mucous phases of encystment and sporular mode of multiplication (Stein).
24. COLPODA PIGERRIMA, Cohn. vol. ii. p. 513, × 600 (Cohn).
25-27. LACRYMARIA COHNII, S. K., vol. ii. p. 518.—At 25 example with the body contorted into a screw-like form, × 250; at 27 an animalcule spherically contracted (Cohn).
28. LACRYMARIA CORONATA, C. & L., vol. ii. p. 518.—The distal extremity only showing the double ciliary wreath, × 350 (Clap. & Lach.).
29-31. TRACHELOCERCA OLOR, Müller sp., vol. ii. p. 515.—29 and 31, Adult animalcules, × 300; 30, distal extremity more highly magnified, displaying oral structure.
32. TRACHELOCERCA PHŒNICOPTERUS, Cohn, vol. ii. p. 516, × 300 (Cohn).
33. TRACHELOCERCA VERSATILIS, Müll. sp., vol. ii. p. 516, × 200.
34. LACRYMARIA LAGENULA, C. & L., vol. ii. p. 517, × 350 (Clap. & Lach.).
35. LAGYNUS ELEGANS, Eng. sp., vol. ii. p. 521, × 300 (Engelmann).

EXPLANATION OF PLATE XXVII. (*continued*).

FIG.
- 36. PHIALINA VERMICULARIS, Ehr., vol. ii. p. 519, × 200 (Ehrenberg).
- 37. PHIALINA VIRIDIS, Ehr., vol. ii. p. 519, × 200 (Ehrenberg).
- 38. TRACHELIUS OVUM, Ehr., vol. ii. p. 522, × 80.
- 39, 40. AMPHILEPTUS ANSER, Ehr., vol. ii. p. 525, × 120 (Ehrenberg).
- 41-44. CHŒNIA TERES, Duj. sp., vol. ii. p. 521.—41, 42, Animalcule in a contracted and extended condition as observed by author, × 200 ; 43, 44, oral region in its expanded and contracted states as delineated by Quennerstedt ; *i* in Fig. 44 represents an ingested food-particle.
- 45, 46. AMPHILEPTUS MELEAGRIS, Ehr. sp., vol. ii. p. 526.—45, Adult animalcule, × 100 (Ehrenberg); 46, showing at *a* an example which has affixed itself to and become encysted upon the branching pedicle of a *Zoothamnium*, the former occupant of which it has first devoured (after D'Udekem).
- 47. TRICHODA PURA, Ehr., vol. ii. p. 535, × 700.
- 48. MENISCOSTOMUM STOMOPTYCHA, Ehr. sp., vol. ii. p. 539, × 200 (Eckhard).
- 49. COLPIDIUM CUCULLUS, Schrank sp., vol. ii. p. 537, × 200.
- 50, 51. PLAGIOPYLA NASUTA (?) Stein, vol. ii. p. 538.—Lateral and ventral aspects, × 200 (Quennerstedt).
- 52. LOXOPHYLLUM MELEAGRIS, Ehr. sp., vol. ii. p. 528, × 110 (Wrzesniowski).
- 53. LOXOPHYLLUM ARMATUM, C. & L., vol. ii. p. 529, × 150 (Clap. & Lach.)
- 54. LEMBADION BULLINUM, Perty, vol. ii. p. 537, × 200.—*m*, Undulating membrane (Clap. & Lach.).
- 55. PLEURONEMA CHRYSALIS, Ehr. sp., vol. ii. p. 543, × 250.
- 56. PLEURONEMA CORONATA, S. K., vol. ii. p. 544, × 200.
- 57, 58. CYCLIDIUM GLAUCOMA, Ehr., vol. ii. p. 544.—57, Normal zooid, × 600; 58, immature example, being as yet deficient of the oral membrane that distinguishes the adult type.
- 59. CYCLIDIUM CITRELLUS, Cohn. sp., vol. ii. p. 545 × 300 (Cohn).
- 60, 61. URONEMA MARINA, Duj., vol. ii. p. 546.—60, Adult, and 61, immature examples, × 600.
- 62, 63. LEMBUS VELIFER, Cohn, vol. ii. p. 547.—62, Adult, and 63, immature example, × 400 (Cohn).
- 64. BÆONIDIUM REMIGENS, Perty, vol. ii. p. 546, × 350 (Fresenius).
- 65, 65*a*. PROBOSCELLA VERMINA, Müll. sp., vol. ii. p. 549.—65, Example as observed and delineated by the author, × 500; 65*a*, illustration of apparently the same species as given in Müller's 'Animalcula Infusoria,' tab. viii. fig. 1, 1786.
- 66, 67. LEMBUS SUBULATUS, S. K., vol. ii. p. 548, × 700.
- 68. AMPHILEPTUS GIGAS, C. & L., vol. ii. p. 526.—Attenuate variety, having the contractile vesicles disposed in linear order along the dorsal border, × 100 ; *tr*, trichocysts (Wrzesniowski).

PLATE XXVII.

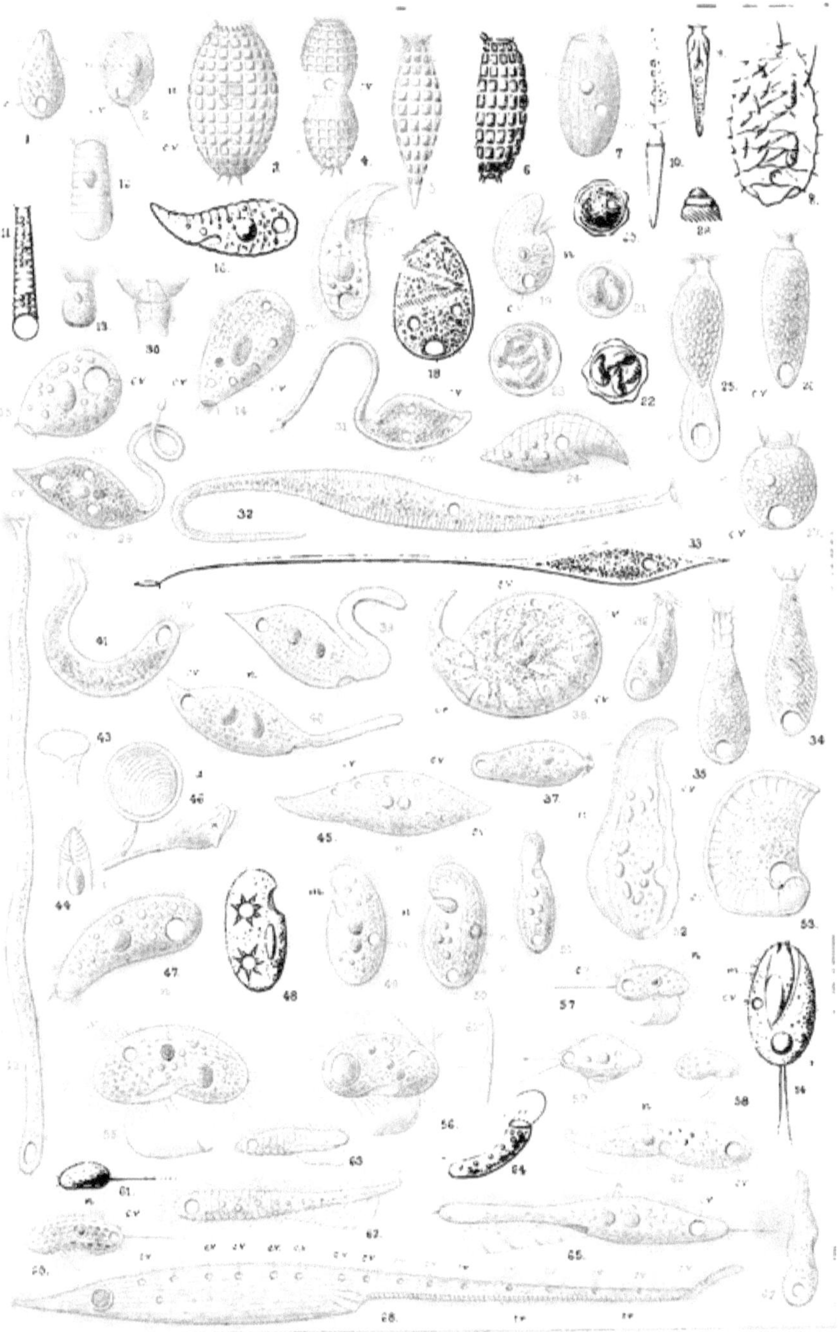

PLATE XXVIII.

EXPLANATION OF PLATE XXVIII.

All the figures in this plate are reproduced from Dr. Joseph Leidy's "Parasites of the Termites," contained in the 'Journal of the Academy of Natural Sciences of Philadelphia,' vol. viii., 1881.

FIG.

1-15. TRICHONYMPHA AGILIS, Leidy, vol. ii. p. 552.—1 and 2, Bilaterally symmetrical animalcules as seen immediately after escaping from the intestine of their host, × 450; at *i* ingested food-particles; 3-5, symmetrical examples, that at 3 having the anterior or distal region of the body spirally involute, those at 4 and 5 with the entire body contorted into respectively shorter and longer helicoidal contours, × 450; 6, bilaterally symmetrical example in a condition of fullest extension, × 450; 7-12, supposed immature examples of the same species, those at 7, 10, 11, and 12, × 666, those at 8 and 9, × 450, the example at Fig. 7 containing numerous spore-like bodies; 13-15, youngest observed examples, × 666 (Leidy).

16-20. PYRSONYMPHA VERTENS, Leidy, vol. ii. p. 554.—16 and 17, distinctly ciliated animalcules, that at 16 containing at *i* numerous ingested particles of wood fibre, *ch* undulating cord-like structure, × 666; 18-20, examples in which no cilia are visible, the periphery of the body presenting a more or less jagged and membranous aspect, × 666; Fig. 19 containing at *i* numerous ingested food-particles (Leidy).

21-24. DINENYMPHA GRACILIS, Leidy, vol. ii. p. 555.—Various more or less contorted examples, Fig. 21 containing a row of spore-like bodies, and 23 and 24 at *i i* ingested food-particles, × 666 (Leidy).

PLATE XXVIII

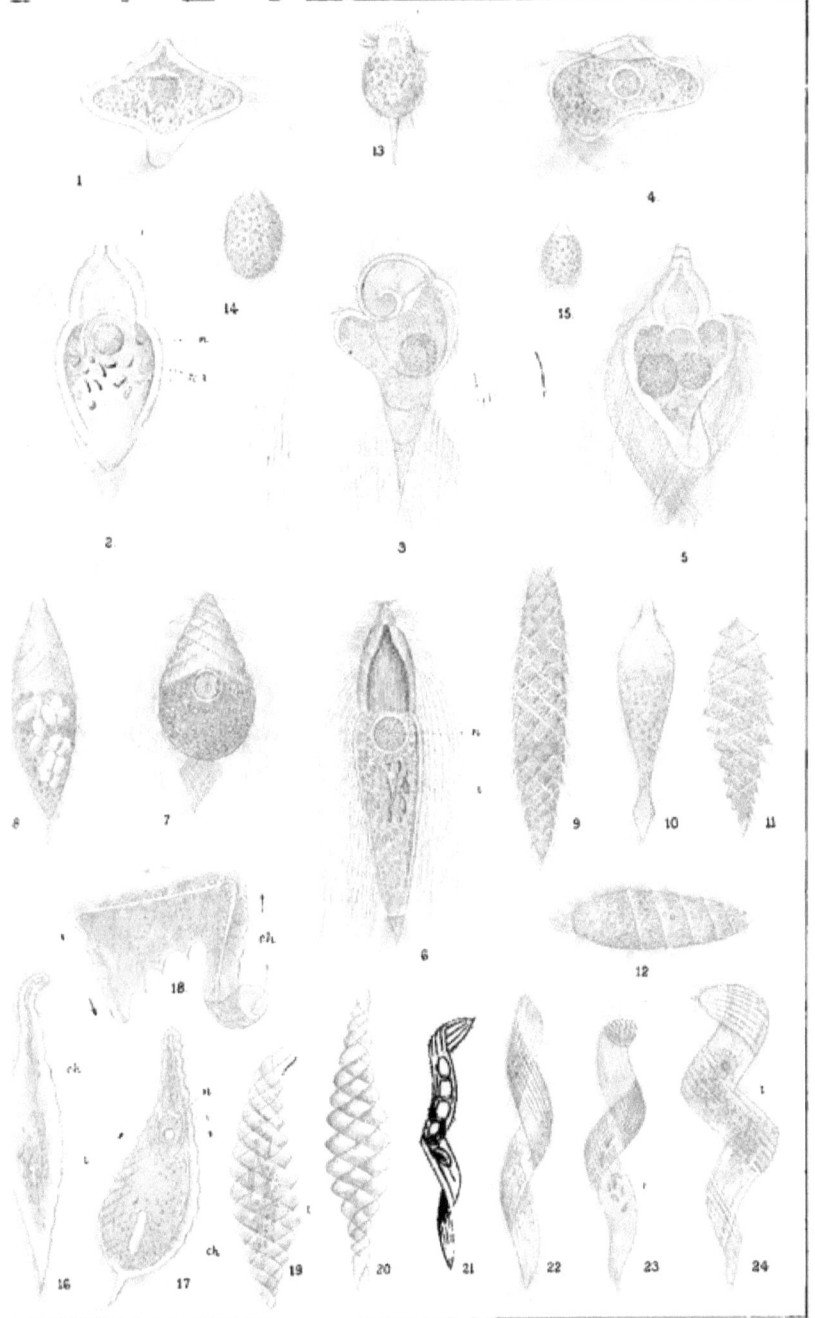

PLATE XXIX.

EXPLANATION OF PLATE XXIX.

FIG.
1–2. BURSARIA TRUNCATELLA, Müller, vol. ii. p. 576.—1, Adult animalcule, × 50; 2, supposed young condition, × 200 (Stein).
3. BALANTIDIUM ENTOZOON, Ehr. sp., vol. ii. p. 577, × 100 (Stein).
4. NYCTOTHERUS CORDIFORMIS, Stein, vol. ii. p. 580, × 150.—*an*, anal aperture (Stein).
5. NYCTOTHERUS VELOX, Leidy. vol. ii. p. 581, × 200 (Leidy).
6–9. METOPUS SIGMOIDES, Clap. & Lach., vol. ii. p. 581.—6 and 8, Extended, 7 and 9, spirally contorted animalcules, × 200 (Stein).
10. PLAGIOTOMA LUMBRICI, Duj., vol. ii. p. 583, × 200 (Stein).
11. CONDYLOSTOMA STAGNALE, Wrz., vol. ii. p. 584, × 150.—*u*, Undulating membrane (Wrzesniowski).
12. CONDYLOSTOMA PATENS, Müll. sp., vol. ii. p. 584, × 150 (Stein).
13, 14. SPIROSTOMUM AMBIGUUM, Ehr., vol. ii. p. 586.—13, An extended, 14, a spirally contorted zooid, × 150 (Stein).
15. BLEPHARISMA UNDULANS, Stein, vol. ii. p. 585, × 150.—*u*, Undulating membrane (Stein).
16, 17. BALANTIDIUM COLI, Malmsten sp., vol. ii. p. 578, × 200.—At 17 an example preparing to divide by transverse fission, and having already developed at *p 2* a second peristome (Stein).
18. LEUCOPHRYS PATULA, Ehr. sp., vol. ii. p. 587, × 150.—*ph*, pharyngeal passage (Stein).
19, 20. FOLLICULINA STYLIFERA, Str. Wright sp., vol. ii. p. 600.—In each instance the anterior region of the lorica is alone represented, the peristome lobes of the contained animalcule in Fig. 19 being fully expanded, while in Fig. 20 they are entirely withdrawn, leaving the stylate appendage only projecting beyond the orifice of the lorica. Dimensions unrecorded (Strethill Wright).
21–28. FOLLICULINA AMPULLA, Müller sp., vol. ii. p. 597.—21 and 22, Entire animalcule and distal region of peristome of var. ACULEATA, Clap. and Lach.; at 24 numerous examples crowded on a shell of *Spirorbis nautiloides*, × 20; 25–27, examples showing the diverse plans of ornamentation of the lorica; 28, normal type in its fully extended state, × 200 (Stein).
29–32. FOLLICULINA PRODUCTA, Str. Wright sp., vol. ii. p. 599.—29, Adult animalcule in its fully extended state, × 200; 30–32, embryonic condition (Strethill Wright).
33–35. FOLLICULINA ELEGANS, Clap. & Lach. sp., vol. ii. p. 598.—33 and 34, Adult animalcules, ventral and lateral aspects, × 200 (Stein); 35, empty lorica showing at *v* valvular elements (Clap. & Lach.).
36. FOLLICULINA BOLTONI, S. K., vol. ii. p. 600, × 200.
37, 38. CHÆTOSPIRA MUELLERI, Lachmann, vol. ii. p. 602, extended and partially retracted states as observed by the author, × 300.
39. FOLLICULINA HIRUNDO, S. K., vol. ii. p. 600, × 200.
40. FOLLICULINA AMPULLA, Müller sp., vol. ii. p. 597, var. VIRIDIS, Str. Wright, × 150 (Strethill Wright).

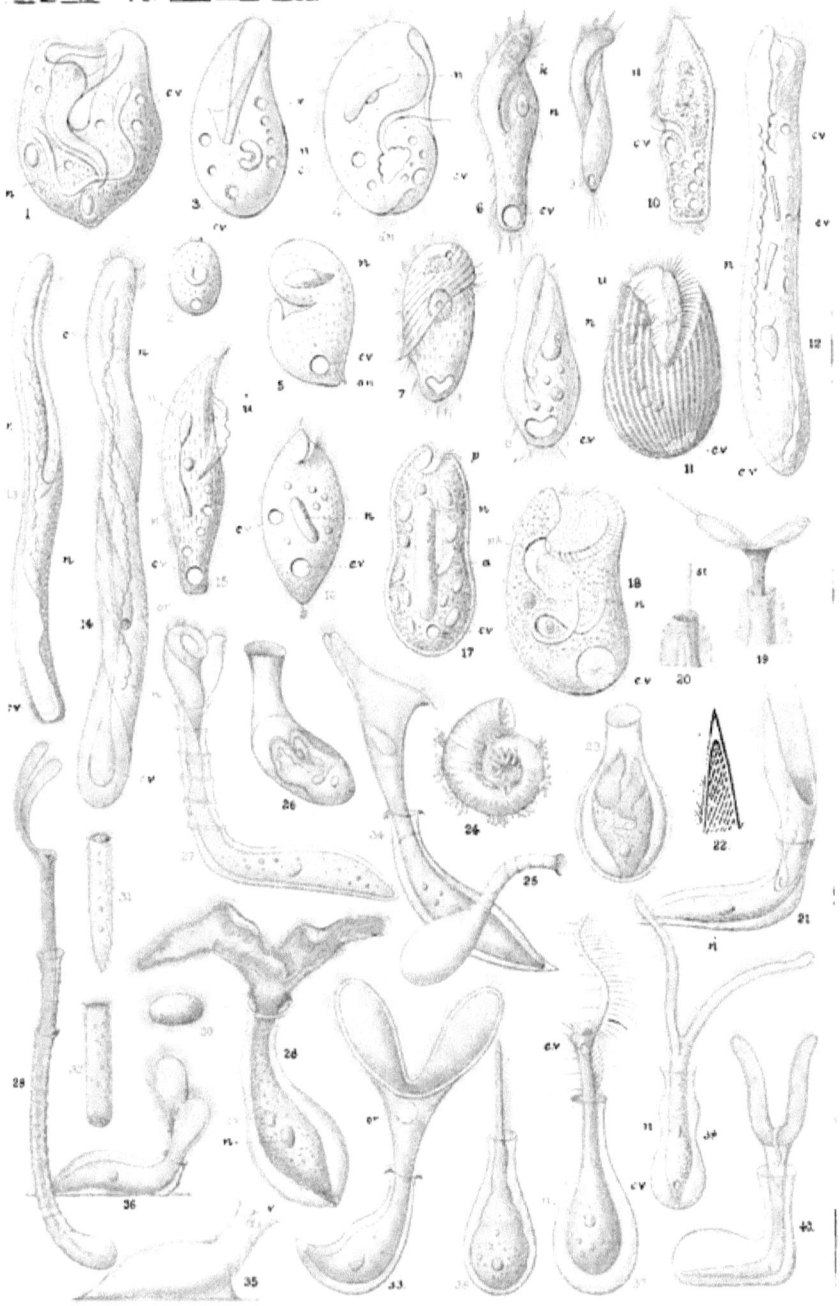

PLATE XXX.

EXPLANATION OF PLATE XXX.

FIG.
1–4. STENTOR IGNEUS, Ehr., vol. ii. p. 594.—1, Attached, fully extended animalcule, × 70; 2 and 3, free-swimming conditions; 4, embryo released by the disintegration of the parent zooid.

5. STENTOR PEDICULATUS, From., vol. ii. p. 596, × 250 (De Fromentel).

6, 7. STENTOR AURICULA, S. K., vol. ii. p. 595, ventral and dorsal aspects, × 120.

8, 9. STENTOR MULTIFORMIS, Stein, vol. ii. p. 595, × 200 (Stein).

10–20. STENTOR POLYMORPHUS, Muller sp., vol. ii. p. 590.—10, Attached, fully extended animalcule, × 120; 11, a social colony, colourless variety, attached to a rootlet of *Anacharis*, which has built up through excretion a common mucilaginous basis, × 15; 12, examples conjugating with their neighbours, as delineated by Balbiani; 13, free-swimming zooid with attenuated posterior region; 14, example with, at *pr*, a laterally developed membranous crest, representing the initial condition of the peristomal ciliary circlet of a future zooid (Ehrenberg); 15, an encysted example (Stein); 16, an animalcule discharging fæcal matter, as observed by the author; 17 and 18, groups of the green variety, natural size, and × 50, after Ehrenberg; 19 and 20, portion of the nucleus or endoplast, showing in the former instance the union of the component granular nodules by an interconnecting cord or funiculus; at 20 an isolated nodule with an enclosed elongate vacuolar structure, × 300 (Stein).

21. STENTOR BARRETTI, Barrett, vol. ii. p. 593.—Fully extended animalcule, × 100.

22–33. STENTOR ROESELII, Ehr., vol. ii. p. 591.—22, Fully extended example affixed within its excreted mucilaginous sheath, × 80; at *pr*, initial condition of a second peristome (Clap. & Lach.); 23, basal or pedal region with pseudopodic peripheral extensions (after Simroth); 24, clavate free-swimming zooid with at *pr* secondary peristome in progress of development (Stein); 25, more advanced condition of fissure, the second peristome, *pr*, having attained a contour and proportions that correspond closely with those of the primary one (Clap. and Lach.); 26, last phase of fissive process, the anterior moiety being now almost completely separated from the posterior one (from a MS. drawing by H. E. Forrest); 27–30, progressive developmental phases, after Claparède and Lachmann; 31, supposed germ-sphere as delineated by Stein; 32, foot or basal region with tuft of longer setose cilia, after Stein; 33, free-swimming semi-contracted animalcule (Stein).

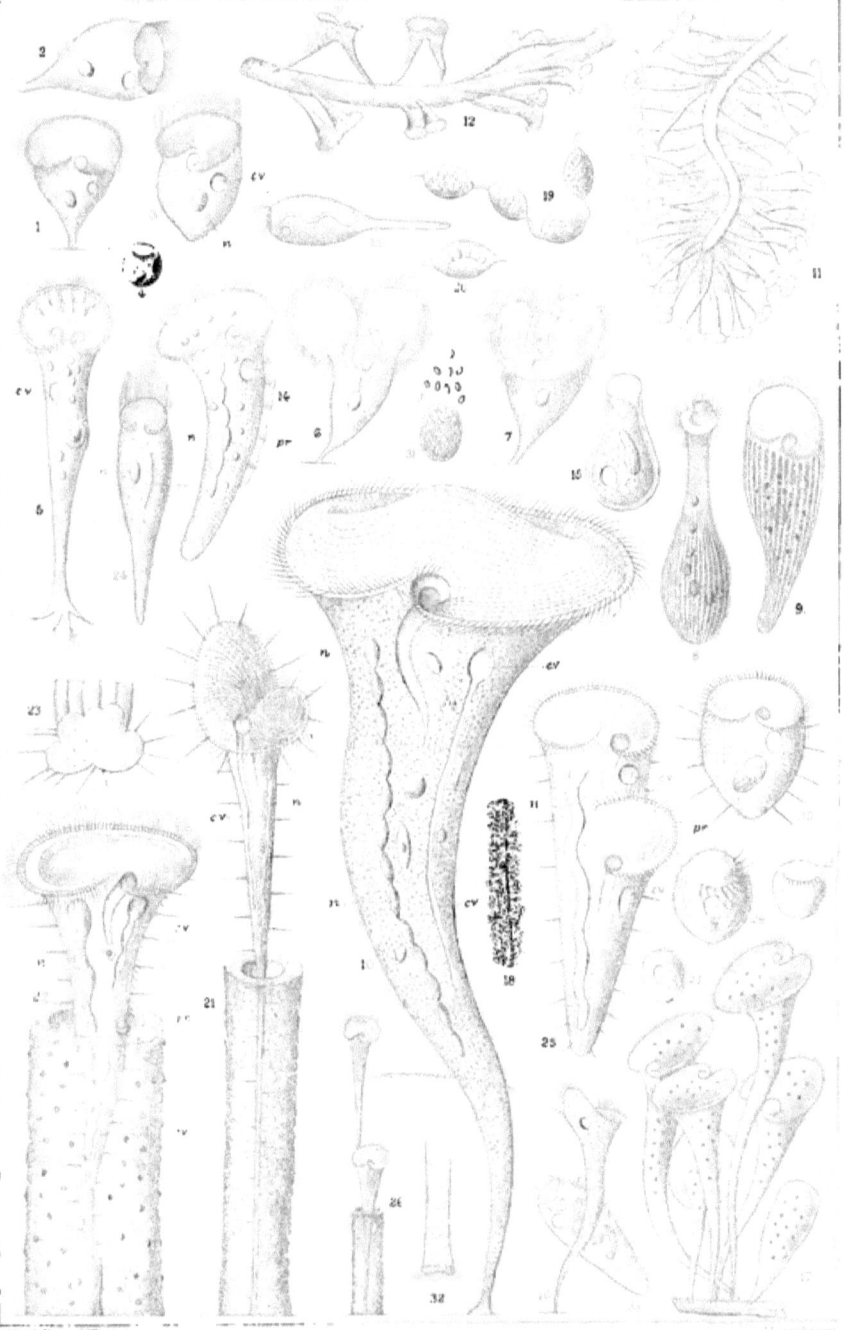

PLATE XXX

PLATE XXXI.

EXPLANATION.

Fig.
1, 2. TINTINNUS EHRENBERGII, C. & L., vol. ii. p. 607, extended and contracted states, × 150 (Claparède & Lachmann).

3. TINTINNUS URNULA, C. & L., vol. ii. p. 609, × 150.—At *pr*, a second peristome in process of development (Clap. & Lach.).

4. TINTINNUS USSOWI, Meresch., vol. ii. p. 609, empty lorica, × 400 (Mereschkowsky).

5. TINTINNUS SUBULATUS, Ehr., vol. ii. p. 605, × 250 (Ehrenberg).

6, 7. TINTINNIDIUM SEMICILIATUM, Sterki sp., vol. ii. p. 612.—At 7 an animalcule isolated from its lorica, viewed in optical section, and showing at *m* outer wreath of furcately branched cirrose cilia or membranellæ, and at *i* inner circlet of ordinary vibratile cilia, × 200 (Sterki).

8. TINTINNIDIUM FLUVIATILE, Stein sp., vol. ii. p. 611, × 200.

9. TINTINNIDIUM MARINUM, S. K., vol. ii. p. 611, × 200 (Ehrenberg).

10. TINTINNUS CINCTUS, C. & L., empty lorica, vol. ii. p. 608, × 250 (Clap. & Lach.).

11. TINTINNUS CAMPANULA, Ehr., × 150.—At *pr*, a second peristome is in course of development, × 150 (Clap. & Lach.).

12. TINTINNUS AMPHORA, C. & L., vol. ii. p. 606, × 150 (Clap. & Lach.).

13. TINTINNUS QUADRILINEATUS, C. & L., vol. ii. p. 607, empty lorica, × 130 (Clap. & Lach.).

14. TINTINNUS ACUMINATUS, C. & L., vol. ii. p. 606, empty lorica, × 200 (Clap. & Lach.).

15. TINTINNUS INQUILINUS, Ehr., vol. ii. p. 604, × 300 (Clap. & Lach.).

16. TINTINNUS MUCICOLA, C. & L., vol. ii. p. 605, × 150 (Clap. & Lach.).

17. TINTINNUS sp.—Empty lorica, × 250 (Clap. and Lach.).

18, 19. TINTINNUS DENTICULATUS, Ehr., vol. ii. p. 607.—18, Empty lorica, × 150; 19, portion of the same, × 800 (Clap. & Lach).

20. TINTINNUS STEENSTRUPII, C. & L., vol. ii. p. 606, × 200 (Clap. & Lach.).

21, 22. TINTINNUS LAGENULA, C. & L., vol. ii. p. 608.—21, normal example, × 300; 22, lorica containing two zooids (Clap. & Lach.).

23. TINTINNUS sp., C. & L., × 150 (Clap. & Lach.).

24. TINTINNUS HELIX, C. & L., vol. ii. p. 608, empty lorica, × 150 (Clap. & Lach.).

EXPLANATION OF PLATE XXXI. (continued).

FIG.
25. TINTINNUS ANNULATUS, C. & L., vol. ii. p. 609, empty lorica, × 150 (Clap. & Lach.).
26. TINTINNUS OBLIQUUS, C. & L., vol. ii, p. 606, × 300 (Clap. & Lach.).
27, 28. VASICOLA CILIATA, Tatem, vol. ii. p. 613.—27, Animalcule contained within its lorica, × 150; 28, free-swimming zooid (Tatem).
29. STROMBIDINOPSIS GYRANS, S. K., vol. ii. p. 614, × 200.
30. TINTINNUS sp., C. & L., empty lorica, × 150 (Clap. & Lach.).
31. TINTINNUS VENTRICOSUS, C. & L., vol. ii. p. 609, empty lorica, × 200 (Clap. & Lach.).
32, 33. CODONELLA GALEA, Hkl., vol. ii. p. 616.—32, Animalcule fully extended beyond the orifice of its lorica, × 200; 33, empty lorica, showing its tesselated character (Haeckel).
34-37. CODONELLA CAMPANELLA, Hkl., vol. ii. p. 618.—34, Fully extended animalcule enclosing centrally numerous ovate germs, × 200; 35, empty lorica; 36, ovate germ, with contained endoplast and contractile vesicle; 37, earlier spore-like condition (Haeckel).
38. CODONELLA ORTHOCERAS, Hkl., vol. ii. p. 616, × 180 (Haeckel).
39-43. TRICHODINOPSIS PARADOXA, C. & L., vol. ii. p. 614.—39, Adult animalcule, profile view, × 200; 40, ventral acetabulum with plicate horny ring, × 300; 41-43, internally developed corneous elements (Clap. & Lach.).
44. URCEOLARIA MITRA, Stein sp., vol. ii. p. 650, × 300 (Clap. & Lach.).
45. TRICHODINA STEINII, C. & L., vol. ii. p. 648.—Acetabulum with denticulate ring, × 250 (Clap. & Lach.).
46, 47. TRICHODINA SCORPÆNA, Robin, vol. ii. p. 649, lateral and ventral views, × 400 (Robin).
48-52. TRICHODINA PEDICULUS, Ehr., vol. ii. p. 646.—48, Examples adherent in various positions to the portion of a tentacle of a *Hydra*; 49, two animalcules in lateral view, × 150; 50, depressed free-swimming zooid; 51, conically contracted animalcule showing at *or* oral aperture, as delineated by Busch; 52, apparent example of conjugation between a large and small zooid, *a*, after Busch.
53. Vol. ii. p. 648. TRICHODINA-like larva, or Trochosphere of Polyzoon, *Alcyonidium gelatinosum*, as delineated in note-book placed at the author's disposal by H. E. Forrest.
54, 55. SCYPHIDIA PHYSARUM, C. & L., vol. ii. p. 658, expanded and contracted zooids, × 200 (Quennerstedt).

PLATE XXXI

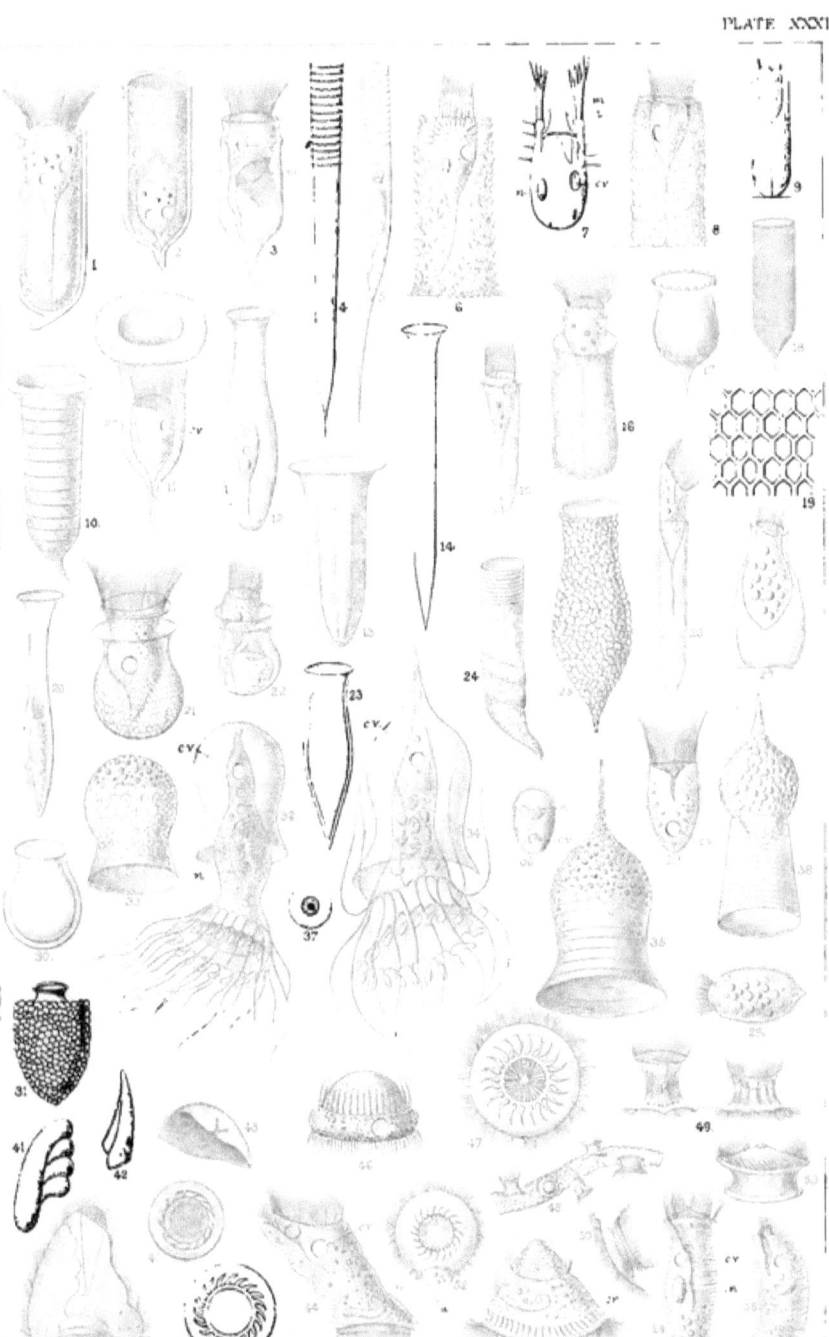

PLATE XXXII.

EXPLANATION OF PLATE XXXII.

FIG.
1–8. ICHTHYOPHTHIRIUS MULTIFILIIS, Fouquet, vol. ii. p. 530.—1, Adult animalcule, × 120 ; 2–5, successive phases of segmentation following upon encystment, and resulting in the subdivision of the entire body-mass into a swarm of minute ciliated germs ; 6, ciliated germ, × 350 ; undeveloped germ with twisted thread-like cuticular extensions of the opposite poles ; 8, oral apparatus viewed from above (Fouquet).
9. PARAMÆCIUM MARINUM, S. K., vol. ii. p. 488, × 250.
10. LEMBUS ELONGATUS, C. & L. sp., vol. ii. p. 549, × 300 (Clap. & Lach.).
11, 12. MARYNA SOCIALIS, Gruber, vol. ii. p. 520.—11, Social colony within branched granular zoothecium, × 60 ; 12, isolated animalcule, × 120 (Gruber).
13. TILLINA MAGNA, Gruber, vol. ii. p. 514, × 190 (Gruber).
14, 15. HAPTOPHRYA GIGANTEA, Maupas, vol. ii. p. 569.—14, Adult animalcule, ventral aspect, × 33 ; 15, acetabulum, × 50 (Certes).
16. LOXOCEPHALUS LURIDUS, Eberhard, vol. ii. p. 489, × 100 (Eberhard).
17. ENCHELYODON ELONGATUS, C. & L., vol. ii. p. 504, × 150 (Clap. & Lach.).
18. METOPIDES CONTORTA, Quenn., vol. ii. p. 583, × 275 (Quennerstedt).
19. OPALINA RANARUM, Purk., vol. ii. p. 559.—Showing temporarily assumed parallel disposition of the cilia as observed by the author, × 100.
20, 21. NYCTOTHERUS GYŒRYANUS, Stein, vol. ii. p. 580.—Ventral and lateral aspects, × 200 (Stein).
22. BALANTIDIUM MEDUSARUM, Meresch., vol. ii. p. 578, × 600 (Mereschkowsky).
23, 24. CALCEOLUS CYPRIPEDIUM, J.-Clk. sp., vol. ii. p. 619.—Ventral and lateral aspects, × 200 (Jas.-Clark).
25, 26. DICTYOCYSTA MITRA, Hkl., vol. ii. p. 623.—26, Empty lorica, × 230 (Haeckel).
27. DICTYOCYSTA TEMPLUM, Hkl., vol. ii. p. 625.—Empty lorica, × 400 (Haeckel).
28. DICTYOCYSTA TIARA, Hkl., vol. ii. p. 626.—Empty lorica, × 400 (Haeckel).'
29–31. DICTYOCYSTA CASSIS, Hkl., vol. ii. p. 624.—29, Empty lorica, showing cribrate structure, × 165 ; 30, extended animalcule, containing germinal bodies g, depending from its lorica, × 165 ; 31, isolated germ, × 300 (Haeckel).
32–34. TORQUATELLA TYPICA, Lankester, vol. ii. p. 621.—32, Animalcule with oral membrane contracted ; 33 and 34, examples, viewed laterally and from above, with this structure variously expanded. Dimensions unrecorded (Ray Lankester).
35–38. HALTERIA GRANDINELLA, Müll. sp., vol. ii. p. 632.—35 and 36, Lateral and ventral aspects, × 600 ; 37, animalcule dividing by transverse fission ; 38, example with its springing-setæ deflected in the act of leaping.
39. HALTERIA VOLVOX, Eichwald, vol. ii. p. 632, × 250 (Claparède & Lachmann).
40. MESODINIUM ACARUS, Stein, vol. ii. p. 635, × 400, as observed by the author.
41. ARACHNIDIUM CONVOLUTUM, S. K., vol. ii. p. 638, × 400.
42, 43. ARACHNIDIUM BIPARTITUM, From. sp., vol. ii. p. 638.—42, Ambulatory, and 43, free-swimming examples, × 400 (De Fromentel).
44. MESODINIUM PULEX, C. & L. sp., vol. ii. p. 636, × 350 (Clap. & Lach.).
45. ACARELLA SIRO, Cohn, vol. ii. p. 636, × 250 (Cohn).
46. STROMBIDIUM CLAPAREDII, S. K., vol. ii. p. 634, × 200.
47. STROMBIDIUM SULCATUM, C. & L., vol. ii. p. 633, × 200 (Clap. & Lach.).
48, 49. ARACHNIDIUM GLOBOSUM, S. K., vol. ii. p. 637.—Examples with tentacles extended and contracted, × 1200.
50–57. DIDINIUM NASUTUM, Müll. sp., vol. ii. p. 639.—50, Normal aspect of natatory animalcule, as observed by the author, × 200 ; 51 and 52, illustrating the seizure and engulfment of a *Paramœcium* as delineated by Balbiani ; 53 and 54, aspects antecedent to and attending transverse fission (Balbiani) ; 55, example of abnormal multiple subdivision as delineated by Eberhard ; 56 and 57, embryonic conditions, after Balbiani.

PLATE XXXII

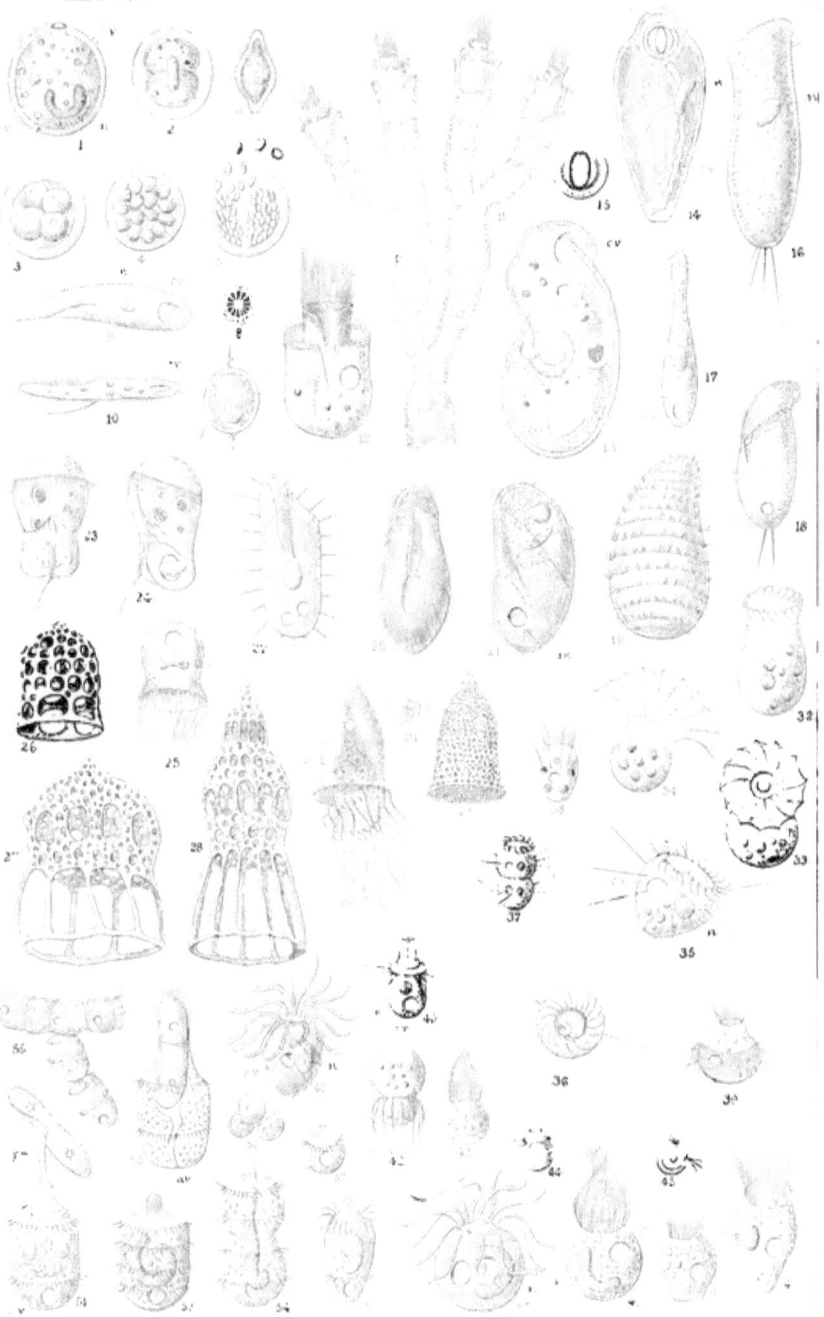

PLATE XXXVI

PLATE XXXVII.

EXPLANATION OF PLATE XXXVII.

FIG.

1–8. ZOOTHAMNIUM ARBUSCULA, Ehr., vol. ii. p. 694.—1, Fully extended adult colony-stock with at *a a* large subspheroidal reproductive zooids, × 100 ; 2, a similar colony-stock, natural size ; 3, colony-stock in a semi-contracted state ; 4, isolated conical zooid, × 350 ; 5, spheroidal reproductive zooid, as delineated by H. E. Forrest ; 6 and 7, successive phases of growth of a reattached reproductive zooid ; 8, ordinary conical zooid with the anterior border plicately contracted.

9–12. ZOOTHAMNIUM DICHOTOMUM, Wright, vol. ii. p. 697.—9, Adult colony-stock, × 50 ; 10, an isolated zooid, × 200 ; 11, colony-stock in semi-contracted state ; 12, scarcer reproductive zooids, × 200.

13, 14. ZOOTHAMNIUM NIVEUM, Ehr., vol. ii. p. 694.—13, Adult colony-stock, × 50 ; 14, isolated zooid, × 200.

15. ZOOTHAMNIUM CIENKOWSKII, Wrz., vol. ii. p. 696, × 400 (Wrzesniowski).

16. ZOOTHAMNIUM PARASITA, Stein, vol. ii. p. 698, × 100 (Stein).

17–19. ZOOTHAMNIUM CARCINI, S. K., vol. ii. p. 700.—17, Adult colony-stock, × 80 ; 18 and 19, expanded and contracted zooids, × 200.

20–24. ZOOTHAMNIUM ALTERNANS, C. & L., vol. ii. p. 695.—20, Adult colony-stock with various sized zooids, as delineated by Greeff, × 125 ; 21, attenuate undeveloped colony-stock, as observed by the author ; 22 and 23, larger reproductive zooids with subcentrally developed supplementary ciliary girdles ; 24, portion of pedicle with finely striate central cord, as observed by the author ; 25, a portion of the same structure, as delineated by H. E. Forrest.

PLATE XXXVII.

PLATE XXXVIII.

EXPLANATION OF PLATE XXXVIII.

FIG.
1–5. EPISTYLIS FLAVICANS, Ehr., vol. ii. p. 702.—Adult erect colony-stock, with at *r r* rosette-like groups of microzooids produced through the repeated segmentation of the ordinary animalcules; at *k k* detached microzooids laterally adherent to and conjugating with ordinary sedentary animalcules, × 100, after Greeff (as originally delineated by Greeff, the right-hand half of the colony is alone portrayed, the left one has been filled by the author, in keeping with the same, for the purpose of symmetry); 2, decumbent colony-stock; 3, zooid after death, the cilia having fallen away, but showing concentric lines of ciliary circlets; 4 and 5, trichocyst-like bodies, as delineated by Greeff, × 600.

6–8. EPISTYLIS PLICATILIS, Ehr., vol. ii. p. 701.—6, Adult colony-stock, × 150; at *a a* commensally attached Flagellata, *Monosiga*; 7 and 8, contracted zooids, × 200.

9. EPISTYLIS UMBILICATA, C. & L., vol. ii. p. 706, × 250 (D'Udekem).

10, 11. EPISTYLIS COARCTATA, C. & L., vol. ii. p. 706.—10, Colony-stock, as delineated by Claparède & Lachmann, × 150; 11, more luxuriant colony, as figured by H. J. Slack; at *a* an apparently encysted zooid.

12–16. EPISTYLIS DIGITALIS, Ehr., vol. ii. p. 704.—12, Adult colony-stock, × 150; 13, isolated zooid, × 300; 14, contracted zooid; 15 and 16, migrant zooids detached from the parent stock, that at 16 progressing on the ground after the manner of a *Trichodina*.

17. EPISTYLIS STEINII, Wrz., vol. ii. p. 708, × 300 (Wrzesniowski).

18. EPISTYLIS NYMPHARUM, Eng., vol. ii. p. 708.—Isolated zooid, × 160 (Engelmann).

19–22. EPISTYLIS ANASTATICA, Linn. sp., vol. ii. p. 701.—19, Colony-stocks attached to limbs of *Cyclops*, × 100; 20, isolated zooid, × 300; 21, contracted animalcule. *Cyclops* with numerous colony-stocks attached, × 10.

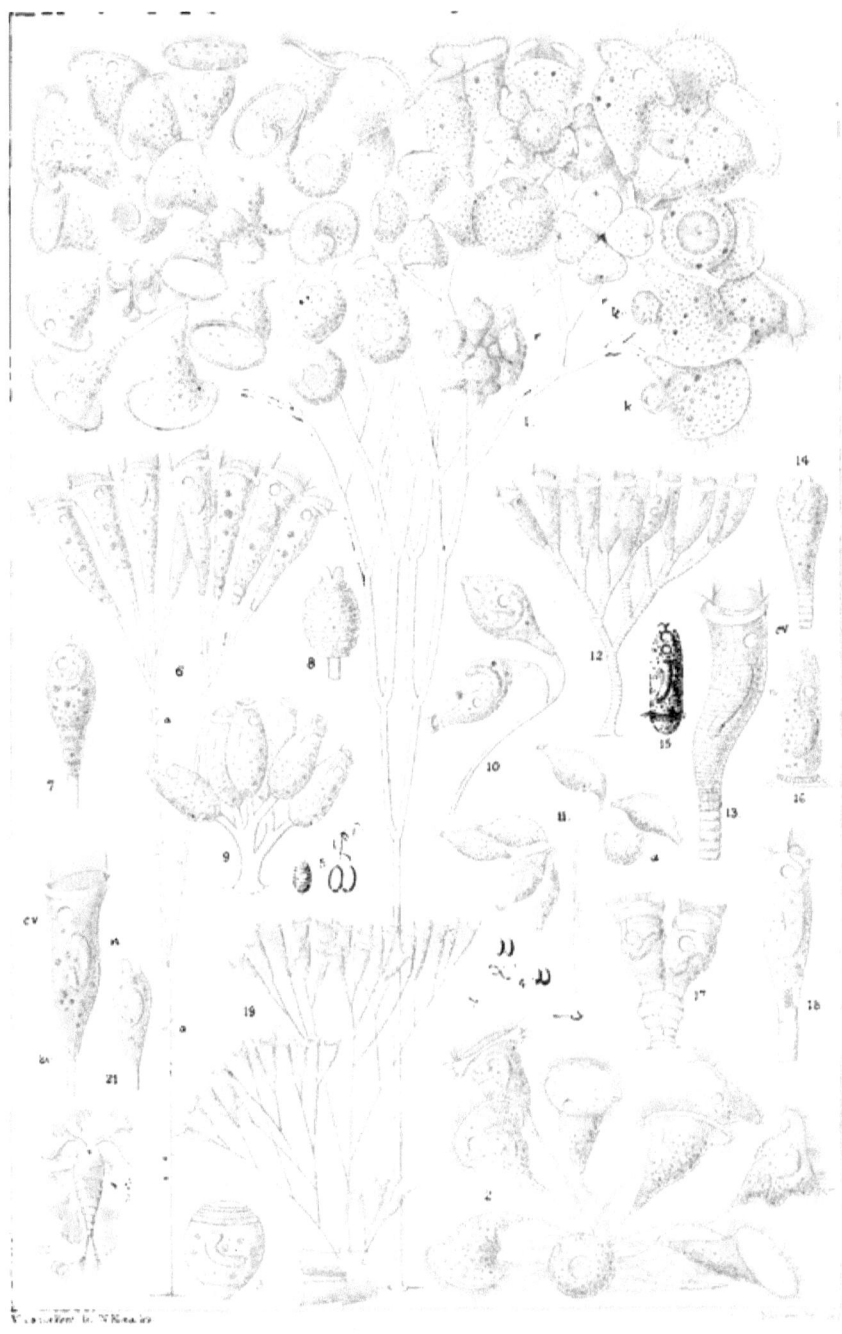

PLATE XXXIX.

EXPLANATION OF PLATE XXXIX.

FIG.
1, 2. EPISTYLIS BRANCHIOPYLA, Perty, vol. ii. p. 705.—1, Fragment of colony-stock, × 200; 2, encysted zooid, × 200 (Stein).

3. EPISTYLIS ARTICULATA, From., vol. ii. p. 707, × 200 (De Fromentel).

4. EPISTYLIS LEUCOA, Ehr., vol. ii. p. 704, × 100 (Ehrenberg).

5. EPISTYLIS BALANORUM, Meresch., vol. ii. p. 709.—Isolated zooid, × 240 (Mereschkowsky).

6. EPISTYLIS GALEA, Ehr., vol. ii. p. 701.—Terminal branch with two zooids, × 120 (Ehrenberg).

7, 8. EPISTYLIS TUBIFICIS, D'Udk., vol. ii. p. 707.—7, Adult colony-stock, × 120; 8, zooid with irregularly developed posterior prolongations (D'Udekem).

9-11. OPERCULARIA HOSPES, From., vol. ii. p. 714.—9, Expanded, 10 and 11, contracted zooids, × 200 (De Fromentel).

12-15. EPISTYLIS PLICATILIS, Ehr., vol. ii. p. 701.—12, Three zooids containing ciliated embryos derived through the disrupture of the endoplast, × 200; at aaa, spout-like apertures whence similar embryos have made their escape; 13, fragment of endoplast enclosing three embryos; 14 and 15, isolated embryos more highly magnified (Claparède & Lachmann).

16. OPERCULARIA MICROSTOMA, Stein, vol. ii. p. 713, × 200 (Stein).

17. OPERCULARIA STENOSTOMA, Stein, vol. ii. p. 712, × 200 (D'Udekem).

18. EPISTYLIS INVAGINATA, C. & L., vol. ii. p. 706.—Isolated zooid, × 200 (Clap. & Lach.)

19. EPISTYLIS CRASSICOLLIS, Stein, vol. ii. p. 705.—Fragment of branch with two zooids, × 200 (Stein).

20. OPERCULARIA LICHTENSTEINII, Stein, vol. ii. p. 712, × 200 (Stein).

21. OPERCULARIA CYLINDRATA, Wrz., vol. ii. p. 713.—Terminal branchlet with two zooids, × 300 (Wrzesniowski).

22, 23. OPERCULARIA NUTANS, Ehr. sp., vol. ii. p. 710.—22, Adult colony-stock, × 300; 23, isolated animalcule with at a a conjugating migrant zooid, as delineated by Ehrenberg.

24-26. OPERCULARIA ARTICULATA, Ehr. sp., vol. ii. p. 711.—24, Portion of adult colony-stock, × 150; 25, isolated zooid; 26, distal extremity showing membranous collar c, and stalked ciliary disc more highly magnified; ee, kidney-shaped corpuscles of undetermined nature (after Stein).

27. OPERCULARIA BERBERINA, Linn. sp., vol. ii. p. 711, × 200 (Stein).

28, 29. VORTICELLA ANNULARIS, Müll., vol. ii. p. 689.—28, Group, natural size, attached to a small *Planorbis*; 29, isolated zooid, × 10 (O. F. Müller).

30. ZOOTHAMNIUM MACROSTYLUM, D'Udk., vol. ii. p. 699.—Dimensions unrecorded (D'Udekem).

31, 32. OPERCULARIA sp., apparently allied to *O. stenostoma*, vol. ii. p. 712.—31, Fragment of colony-stock; 32, isolated zooid, as figured by Henry Baker, 'Employment for the Microscope,' 1785, pl. xiii. figs. 13 and 14, dimensions unrecorded.

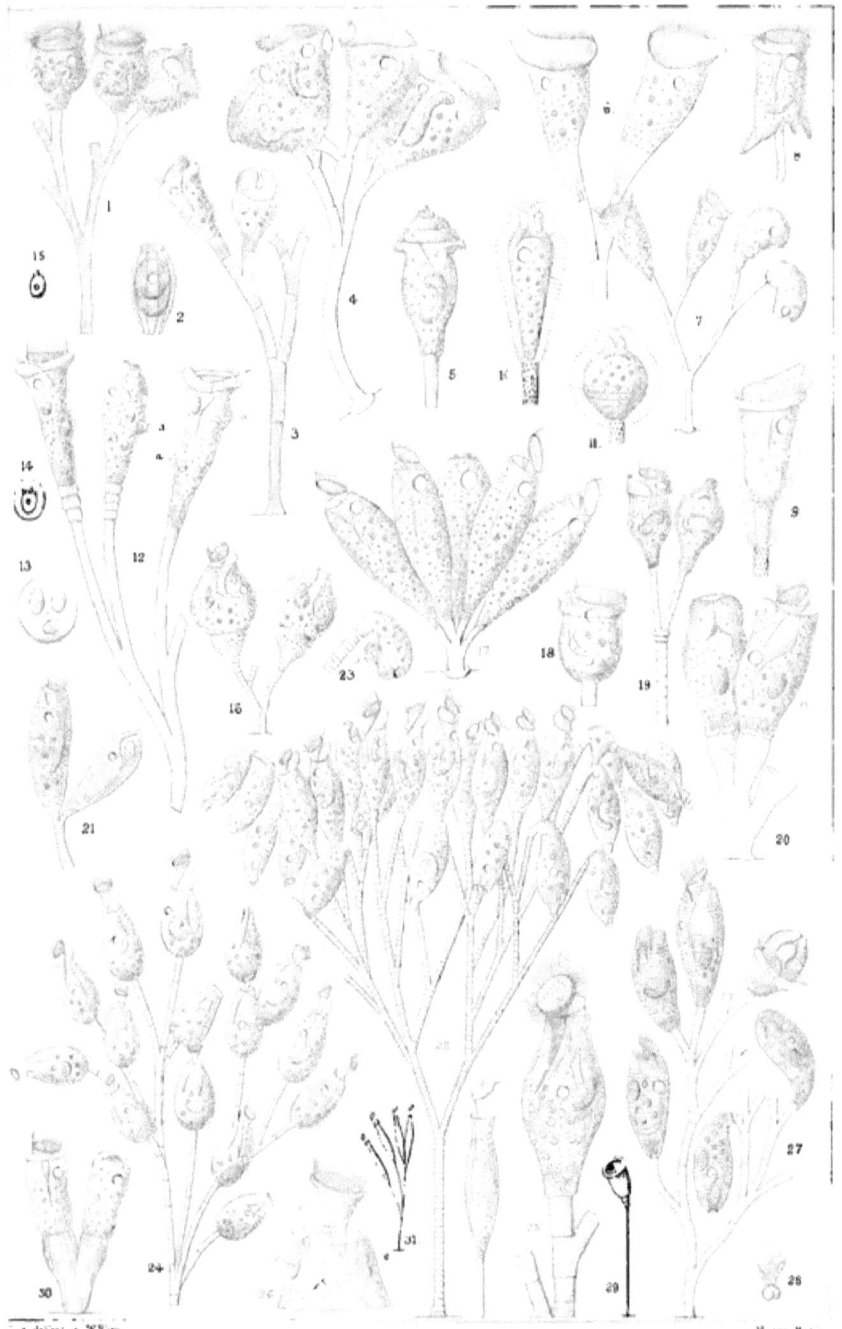

PLATE XXXIX

PLATE XL.

EXPLANATION.

FIG.
1. VAGINICOLA CRYSTALLINA, Ehr., vol. ii. p. 715.—Lorica containing two zooids, × 200.
2. STYLOCOLA STRIATA, From., vol. ii. p. 730, × 300 (De Fromentel).
3. VAGINICOLA GLOBOSA, D'Udk., vol. ii. p. 716, dimensions unrecorded (D'Udekem).
4, 5. THURICOLA VALVATA, Wright sp., vol. ii. p. 718.—Loricæ with two zooids in an expanded and contracted state; v, valvular structure, × 200 (Str. Wright).
6-8. THURICOLA FOLLICULATA, Müll. sp., vol. ii. p. 718.—6, Fully extended zooid with at vv lateral edges of pectinate valve, × 200; 7 and 8, distal region of lorica showing pectinate structure of valve, × 400.
9, 10. COTHURNIA IMBERBIS, Ehr., vol. ii. p. 720.—9, Lorica containing two zooids, that at a in a contracted condition, having developed a subcentral ciliary girdle preparatory to migration, × 200; 10, detached migratory zooid (Greeff).
11. COTHURNIA PONTICA, Meresch., vol. ii. p. 725.—Empty lorica, × 150 (Mereschkowsky).
12. COTHURNIA GRACILIS, S. K., vol. ii. p. 724.—Zooid contracted, × 250.
13-15. THURICOLA OPERCULATA, Gruber sp., vol. ii. p. 719.—13 and 14, Expanded zooids, showing at op valve-like operculum, and at l retractile ligament that connects the same with the body at the bottom of the lorica; 15, basal region with the operculum and retractile ligament diagrammatically represented (Gruber).
16. PYXICOLA SOCIALIS, Gruber sp., vol. ii. p. 728.—Distal region of expanded zooid with at m collar-like membrane, dimensions unrecorded (Gruber).
17, 18. COTHURNIA COMPRESSA, C. & L., vol. ii. p. 722.—Front and lateral aspects, × 170 (Claparède & Lachmann).
19. COTHURNIA HAVNIENSIS, Ehr., vol. ii. p. 720, × 100 (Ehrenberg).
20, 21. COTHURNIA PATULA, From., vol. ii. p. 722.—Lorica with two zooids in their expanded and contracted states, × 200 (De Fromentel).
22. COTHURNIA COHNII, S. K., vol. ii. p. 723, × 400 (Cohn).
23. COTHURNIA PUPA, Eichw., vol. ii. p. 724, dimensions unrecorded (Eichwald).
24, 25. COTHURNIA SIEBOLDII, Stein, vol. ii. p. 720.—Ventral and lateral aspects, × 150 (Stein).
26. COTHURNIA ASTACI, Stein, vol. ii. p. 721, × 150 (Stein).
27. COTHURNIA CURVA, Stein, vol. ii. p. 721, × 300 (Stein).

EXPLANATION OF PLATE XL.—*continued.*

FIG.
28, 29. PYXICOLA AFFINIS, S. K., vol. ii. p. 727.—Expanded and contracted zooids ; *op*, operculum, × 250.
30, 31. PYXICOLA SOCIALIS, Gruber sp., vol. ii. p. 728.—30, Social group, × 90 ; 31, isolated zooid, × 250 ; *op*, operculum (Gruber).
32. PACHYTROCHA COTHURNOIDES, S. K., vol. ii. p. 729.—Extended zooid with at *op* fleshy operculum, × 500.
33, 34. PLATYCOLA DECUMBENS, Ehr. sp., vol. ii. p. 731.—Dorsal and lateral aspects, the lorica at 33 containing two zooids, × 280.
35. PLATYCOLA LONGICOLLIS, S. K., vol. ii. p. 732, × 300 (De Fromentel).
36-38. LAGENOPHRYS VAGINICOLA, Stein, vol. ii. p. 733.—36, Two zooids attached to a branched hair of *Canthocamptus minutus*, × 350 ; 37, example with at *g* posteriorly separated germ ; 38, membranous valve of oral region.
39. PYXICOLA OPERCULIGERA, S. K., vol. ii. p. 725.—Extended zooid with at *op* discoidal operculum, × 250.
40. PYXICOLA CARTERI, S. K., vol. ii. p. 729, × 400.—*op*, Operculum.
41. PYXICOLA PYXIDIFORMIS, D'Udk. sp., vol. ii. p. 726.—Extended zooid with at *op* lid-like operculum, × 150 (D'Udekem).
42. PLATYCOLA STRIATA, From., vol. ii. p. 732, × 300 (De Fromentel).
43. PLATYCOLA DILATATA, From., vol. ii. p. 731, × 300 (De Fromentel).
44-46. LAGENOPHRYS AMPULLA, Stein, vol. ii. p. 733.—44, Extended zooid with at *g g* detached ciliated germs, × 300 ; 45 and 46, detached germs more highly magnified, showing hypotrichous plan of ciliation (Stein).
47. LAGENOPHRYS NASSA, Stein, vol. ii. p. 733.—Retracted zooid, × 300 (Stein).

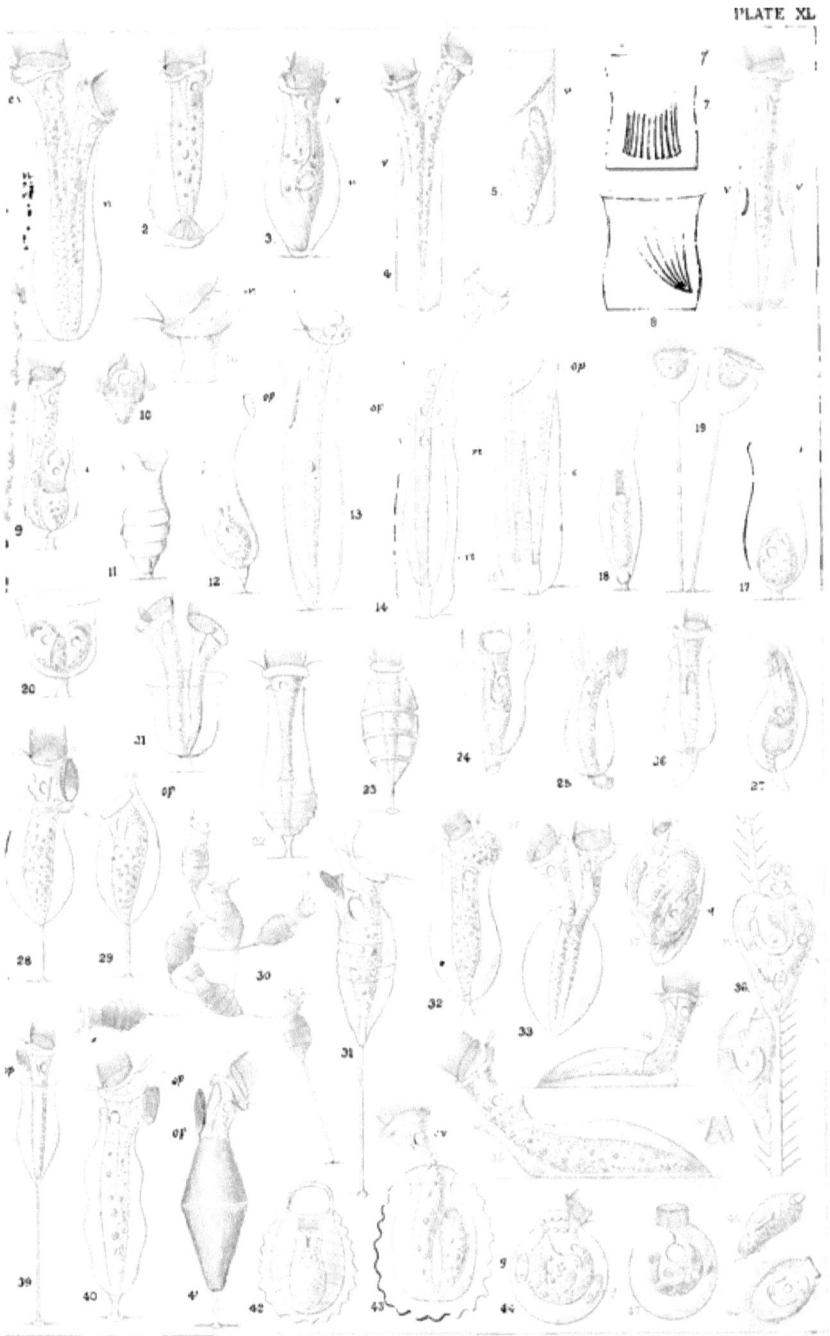

PLATE XL

PLATE XLI.

EXPLANATION OF PLATE XLI.

Fig.

1–9. OPHRYDIUM VERSATILE, Müll. sp., vol. ii. p. 735.—1 and 2, Colony-stocks, natural size (after Ehrenberg) ; 3, portion of a similar colony-stock, as seen in optical section slightly magnified ; 4, fragment of a colony-stock with fully extended zooids, showing at *a a* the thread-like branching pedicle with its surrounding common gelatinous matrix or zoocytium *z*, × 100 (after Wrzesniowski) ; 5 and 6, extended and contracted zooids, × 200 (after Stein) ; 7, basal region of three zooids of a variety in which the relatively short pedicles are distinctly annulate at their distal ends (Wrzesniowski) ; 8, fragments of a colony-stock treated with hæmatoxylin and osmic acid, showing at *a* the urceolate cavities in the zoocytium into which the zooids retreat ; *b*, a retracted zooid ; and *c*, the pedicle ; the lines at *d* indicate the boundaries of the zoocytial element exuded by the individual zooids, × 150 (Wrzesniowski) ; 9, a detached free-swimming zooid having a posteriorly developed supplementary girdle of cilia.

10–17. OPHRYDIUM EICHORNII, Ehr., vol. ii. p. 737.—10, a moderate-sized colony-stock with fully extended zooids, as observed by the author, × 50 ; 11, a fragment of the same stock with, at *a, b*, contracted zooids, × 150 ; 12, a zooid with posteriorly developed ciliary girdle preparing to enter upon the free-swimming state ; 13, a zooid exhibiting the rare phenomenon of subdivision by transverse fission, the line at *tr* indicating the region of separation ; 14, a detached free-swimming zooid ; 15 and 16, whole and portion of a normal zooid × 400 (as delineated by Wrzesniowski) ; *or*, oral entrance ; *v*, vestibular fossa ; *ph*, pharynx ; *œ*, œsophageal tube ; 17 and 18, outlines of the contractile vesicle.

19–21. OPHRYDIUM SESSILE, S.K., vol. ii. p. 738.—19, colony-stock attached to vegetable filament, natural size ; 20, a colony-stock with fully extended zooids, × 100 ; 21, a similar stock with the zooids retracted within their common gelatinous zoocytium.

22, 23. OPHIONELLA PICTA, S.K., vol. ii. p. 734.—22, extended, and 23, contracted zooid, × 300.

PLATE XLI

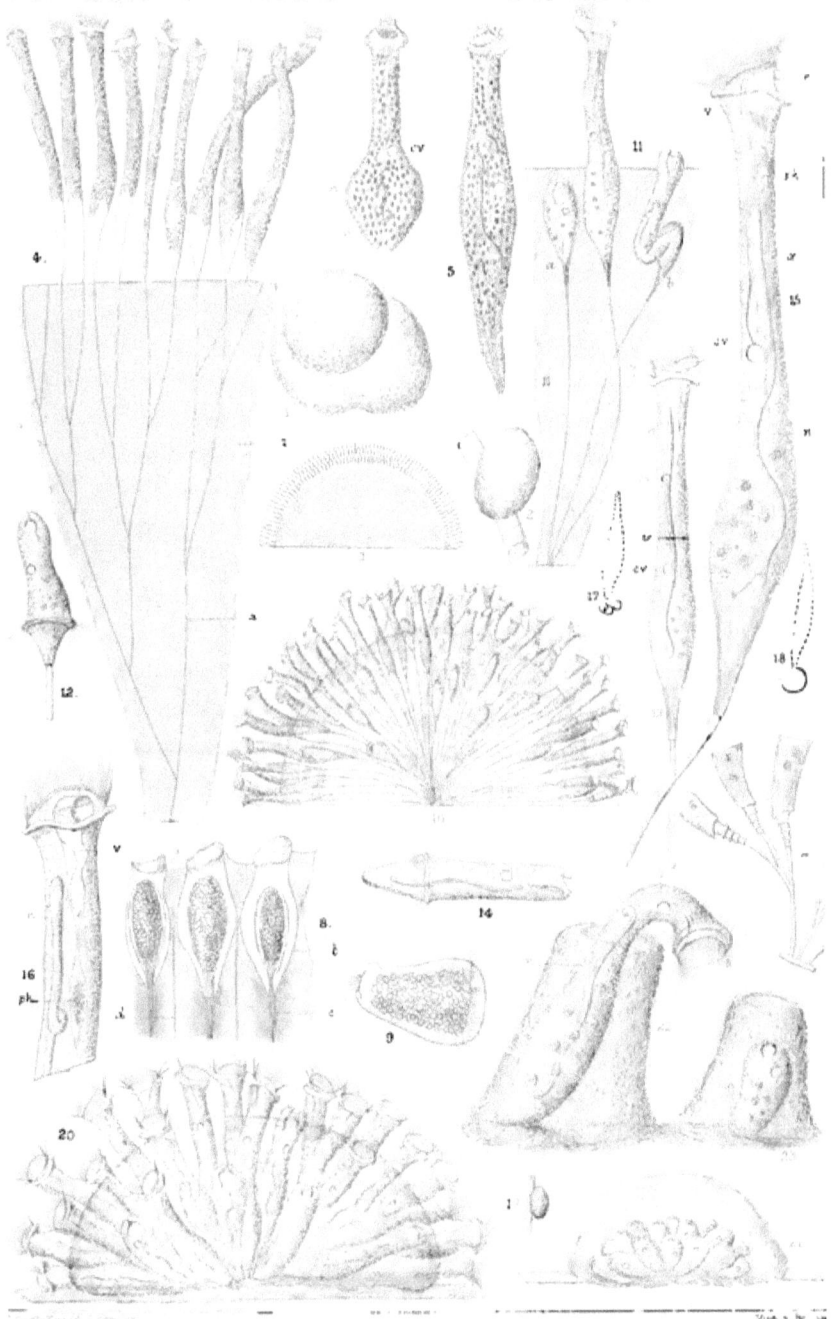

PLATE XLII.

EXPLANATION.

FIG.
1-3. LOXODES ROSTRUM, Ehr., vol. ii. p. 748.—1, zooid as viewed superficially, showing cuticular striæ, marginal setæ, and endoplastular spherules; *ph*, sickle-shaped pharyngeal tube, × 200; 2, a similar zooid more deeply focussed, showing the reticulate or cancellate structure of the internal parenchyma; *v*, vacuolar spaces; *r*, refringent corpuscles; 3, indurated pharynx, × 300 (after Wrzesniowski).

4. LITONOTUS VARSAVIENSIS, Wrz., vol. ii. p. 744, × 500.—*tr*, trichocysts (Wrzesniowski).

5-11. LITONOTUS FASCIOLA, Ehr. sp., vol. ii. p. 743.—5, lateral view, and 6, ventral aspect, × 500 (Wrzesniowski); 7, example abnormally dilated with ingested food; 8-11, successive developmental phases of a single zooid (as observed by the author).

12, 13. LITONOTUS WRZESNIOWSKII, S. K., vol. ii. p. 742.—12, Lateral, and 13, ventral aspects, × 400; *tr*, trichocysts.

14, 15. PHASCOLODON VORTICELLA, Stein, vol. ii. p. 746.—Ventral and lateral aspects, × 250 (Stein).

16-22. CHILODON CUCULLULUS, Müll. sp., vol. ii. p. 746.—16 and 17, Ventral and lateral aspects, × 200 (Stein); 18, example dividing by transverse fission; 19, conjugation of two independent zooids (Stein); 20, zooid with abnormally developed pharyngeal tube, drawn from a preserved example supplied to the author by Mr. Charles Stewart; 21 and 22, pharyngeal rod-fascicles, showing extended and contracted conditions during the ingestion of a diatom frustule.

23. OPISTHODON NIEMECCENSIS, Stein, vol. ii. p. 750.—Ventral aspect, × 150 (Stein).

24-26. IDUNA SULCATA, C. & L., vol. ii. p. 752.—Dextral, sinistral, and ventral aspects, × 175 (Claparède & Lachmann).

27-30. DYSTERIA ARMATA, Huxley, vol. ii. p. 752.—27, Sinistral aspect, × 250; 28-30, corneous elements of pharyngeal apparatus further enlarged (after Huxley).

31-33. CYPRIDIUM LANCEOLATUM, C. & L. sp., vol. ii. p. 754.—31 and 32, Dextral and sinistral aspects, × 350; 33, ventral view of posterior region, showing the union of the lateral valves, and at *st* caudal style (Claparède & Lachmann).

EXPLANATION OF PLATE XLII. (*continued*).

FIG.
34. CYPRIDIUM SPINIGERUM, C. & L., vol. ii. p. 754.—Dorsal aspect, × 300 (Claparède & Lachmann).
35, 36. ÆGYRIA MONOSTYLA, Ehr. sp., vol. ii. p. 755.—Dextral and sinistral views, × 200 (Stein).
37, 38. HUXLEYA CRASSA, C. & L., vol. ii. p. 758, × 300 (Claparède & Lachmann).
39, 40. ÆGYRIA ANGUSTATA, C. & L., vol. ii. p. 755.—Dextral and ventral aspects, × 300 (Claparède & Lachmann).
41, 42. CHLAMYDODON MNEMOSYNE, Ehr., vol. ii. p. 750.—Ventral and lateral aspects, × 250 (Stein).
43, 44. ÆGYRIA OLIVA, C. & L., vol. ii. p. 756.—Dextral and sinistral aspects, × 250; *e*, eye-like pigment-speck (Claparède & Lachmann).
45. CYPRIDIUM ACULEATUM, C. & L. sp., vol. ii. p. 754.—Sinistral aspect, × 250 (Claparède & Lachmann).
46. TRICHOPUS DYSTERIA, C. & L., vol. ii. p. 758, × 200 (Claparède & Lachmann).
47, 48. TROCHILIA SIGMOIDES, Duj., vol. ii. p. 757.—47, as figured by Claparède & Lachmann, under the title of *Huxleya sulcata*, × 500; 48, after Dujardin.
49, 50. SCAPHIDIODON NAVICULA, Stein, vol. ii. p. 750.—Ventral and dorsal aspects, × 240 (Stein).
51, 52. TROCHILIA PALUSTRIS, Stein, vol. ii. p. 757.—Dorsal and ventral aspects, × 400 (Stein).

PLATE XLII

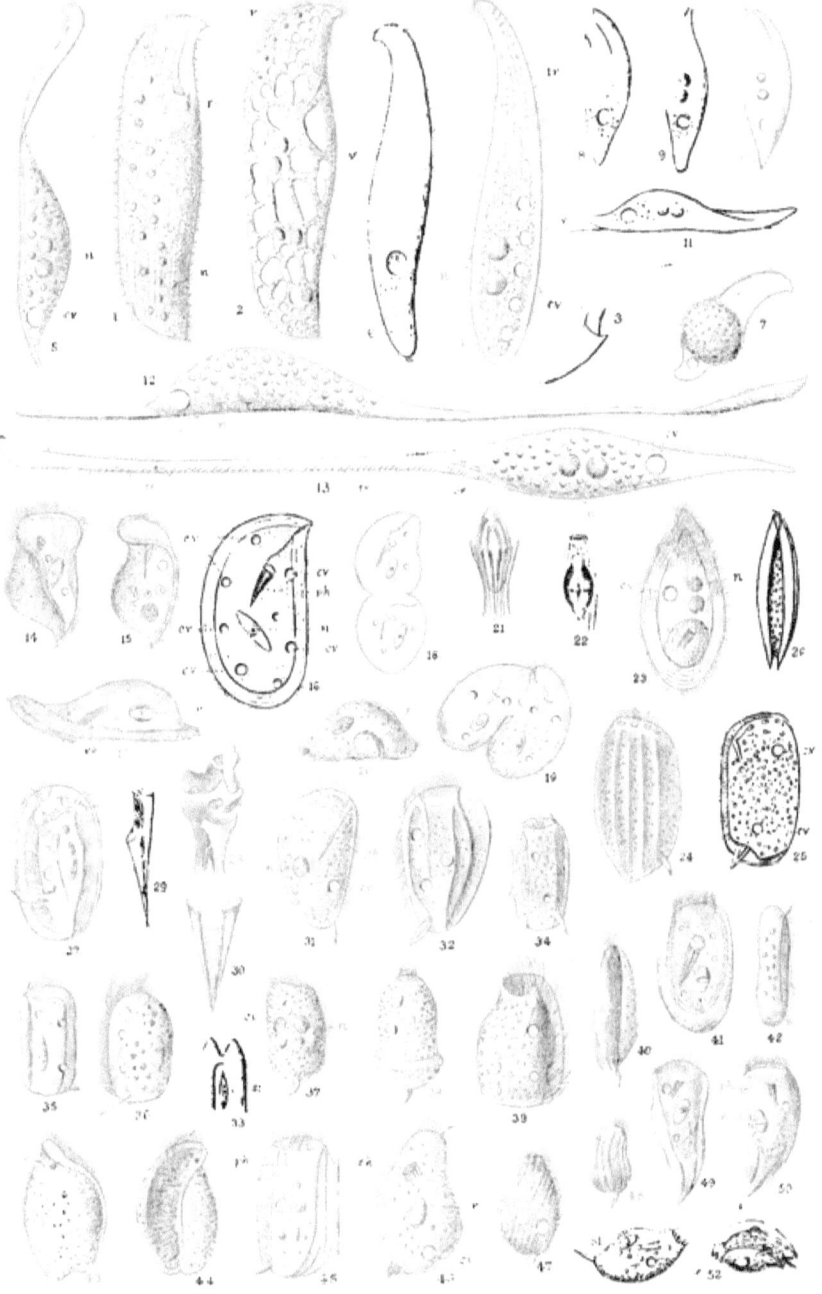

PLATE XLIII.

EXPLANATION OF PLATE XLIII

FIG.
1-3. PSILOTRICHA ACUMINATA, Stein, vol. ii. p. 762.—1, Ventral, 2, dorsal, and 3, lateral aspects, × 300 (Stein).

4-5. KERONA POLYPORUM, Ehr., vol. ii. p. 763.—Ventral and dorsal aspects, × 200 (Stein).

6-8. UROSTYLA GRANDIS, Ehr., vol. ii. p. 765.—6, Elongate example in ventral aspect ; 7, shorter zooid, dorsal view, × 100 ; 8, encysted zooid (Stein).

9-10. UROLEPTUS MOBILIS, Eng., vol. ii. p. 781.—Extended and contracted examples, × 150 (Engelmann).

11. HOLOSTICHA MYSTACEA, Stein sp., vol. ii. p. 769.—Ventral aspect, × 150 (Stein).

12. AMPHISIA PERNIX, Wrz. sp., vol. ii. p. 768.—Ventral aspect, × 300 (Wrzesniowski).

13. ONYCHODROMUS GRANDIS, Stein, vol. ii. p. 766.—Ventral aspect, × 150 (Stein).

14. UROLEPTUS MUSCULUS, Müll. sp., vol. ii. p. 779.—Ventral aspect, × 150 (Stein).

15. AMPHISIA GIBBA, Müll. sp., vol. ii. p. 767.—Ventral aspect, × 200 (Stein).

16. GASTROSTYLA STEINII, Eng., vol. ii. p. 784.—Ventral aspect, × 200 (Engelmann).

17. HOLOSTICHA RUBRA, Ehr. sp., vol. ii. p. 770.—Ventral view, × 300 (Cohn).

18. UROLEPTUS VIOLACEUS, Stein, vol. ii. p. 781.—Ventral aspect, × 200 (Stein).

19, 20. HOLOSTICHA FLAVA, Cohn sp., vol. ii. p. 769.—19, Lateral, and 20, ventral aspects, × 250 (Cohn).

21. UROLEPTUS PISCIS, Müll. sp., vol. ii. p. 780.—Ventral aspect, × 100 (Stein).

22. PLEUROTRICHA ECHINATA, C. & L. sp., vol. ii. p. 783.—Ventral aspect, × 300 (Claparède & Lachmann).

23, 24. EPICLINTES RETRACTILIS, C. & L. sp., vol. ii. p. 774.—23, Extended, and 24, contracted conditions, ventral view, × 500 (Claparède & Lachmann).

25. PLAGIOTRICHA AFFINIS, Stein sp., vol. ii. p. 772.—Ventral aspect, × 300 (Stein).

26, 27. PLEUROTRICHA LANCEOLATA, Ehr. sp., vol. ii. p. 783.—Ventral aspect, × 150 ; 27, encysted zooid (Stein).

28-30. EPICLINTES AURICULARIS, C. & L. sp., vol. ii. p. 773.—28, Ventral view ; 29, repent animalcule in lateral aspect, × 200 (after Claparède & Lachmann ; 30, anterior region more highly magnified, as delineated by Mereschkowsky.

31, 32. EPICLINTES RADIOSA, Quenn. sp., vol. ii. p. 774.—31, Dorsal, and 32, lateral aspects, × 300 (Quennerstedt).

33. UROLEPTUS RATTULUS, Stein, vol. ii. p. 780.—Ventral aspect, × 150 (Stein).

34. PLAGIOTRICHA STRENUA, Eng. sp., vol. ii. p. 772.—Ventral view, × 180 (Engelmann).

35, 36. OPISTHOTRICHA PARALLELA, Eng. sp., vol. ii. p. 785.—35, Dorsal, and 36, ventral aspects, × 150 (Engelmann).

PLATE XLIII

PLATE XLIV.

EXPLANATION OF PLATE XLIV.

FIG.

1, 2. STICHOTRICHA SECUNDA, Perty, vol. ii. p. 776.—1, Zooid projecting from its mucilaginous sheath ; 2, isolated animalcule in ventral aspect, × 250 (Stein).

3. STICHOTRICHA ACULEATA, Wrz., vol. ii. p. 777.—Ventral aspect × 500 (Wrzesniowski).

4-8. SCHIZOSIPHON SOCIALIS, Gruber sp., vol. ii. p. 778.—4, pendent dichotomously branching colony-stock or zoothecium with terminally enclosed zooids, × 60 ; 5, an isolated animalcule, × 240 ; 6, an example dividing by transverse fission within its tube ; 7, termination of tube temporarily occupied by two zooids ; 8, a colony-stock, natural size (Gruber).

9, 10. STICHOTRICHA REMEX, Hudson, vol. ii. p. 777.—9, Animalcules projecting in various positions from their cylindrical sheaths, × 40 ; 10, an isolated zooid, × 150 (C. S. Hudson).

11. STICHOTRICHA CORNUTA, C. & L., vol. ii. p. 776.—Zooid in ventral aspect, × 400 (Claparède & Lachmann).

12. OXYTRICHA TUBICOLA, Gruber, vol. ii. p. 789.—Zooid enclosed within its tube, × 200 (Gruber).

13, 14. STICHOCHÆTA PEDICULIFORMIS, Cohn, vol. ii. p. 775.—13, Ventral, and 14, lateral aspects, × 500 (Cohn).

15, 16. AMPHISIA GIBBA var. CRASSA, C. & L. sp., vol. ii. p. 768.—Ventral, and lateral aspects, × 200 (Claparède & Lachmann).

17, 18. OXYTRICHA SCUTELLUM, Cohn, vol. ii. p. 788.—Extended and contracted zooids, × 360 (Cohn).

19, 20. OXYTRICHA ÆRUGINOSA, Wrz., vol. ii. p. 787.—19, Ventral and 20, lateral aspects, × 150 (Wrzesniowski).

21. HOLOSTICHA OCULATA, Meresch. sp., vol. ii. p. 770.—Ventral aspect, dimensions unrecorded (Mereschkowsky).

22. EUPLOTES HARPA, Stein, vol. ii. p. 799.—Ventral aspect, × 150 (Stein).

23-25. EUPLOTES PATELLA, Ehr., vol. ii. p. 798.—23, Dorsal, and 24, ventral aspects, × 200 (after Stein) ; 25, lateral aspect (Dujardin).

26-29. EUPLOTES CHARON, Müll. sp., vol. ii. p. 799.—26, Ventral view, × 300 ; 27, example dividing by transverse fission ; 28, encysted zooid (Stein).

30, 31. STYLOPLOTES APPENDICULARIS, Ehr. sp., vol. ii. p. 800.—Dorsal and ventral aspects, × 300 (Stein).

PLATE XLV.

EXPLANATION OF PLATE XLV.

FIG.
1. STYLONYCHIA MYTILUS, Ehr., vol. ii. p. 790.—Diagrammatic plan of oral ciliary system: *ad*, adoral cilia or membranette; *præ*, præoral, and *en*, endoral cilia; *u*, undulating membrane; *fr*, frontal style (Sterki).
2. GASTROSTYLA, sp. (Sterki).—Diagrammatic optic section showing disposition of the several ciliary systems; *d*, dorsal, and *m*, marginal setæ; *par*, paroral cilium; the other lettering as in the preceding figure.
3-5. OXYTRICHA PELLIONELLA, Müll. sp., vol. ii. p. 786.—3, Ventral aspect, × 400; 4, diagrammatic longitudinal section: *ad*, adoral, *fr*, frontal, *v*, ventral, *an*, anal, and *d*, dorsal cilia or setæ (after Sterki); 5, ventral aspect (after Stein).
6. ACTINOTRICHA SALTANS, Cohn, vol. ii. p. 790, × 360 (Cohn).
7. OXYTRICHA FALLAX, Stein, vol. ii. p. 787.—Ventral aspect, × 170 (Stein).
8, 9. OXYTRICHA PLATYSTOMA, Ehr. sp., vol. ii. p. 787.—Ventral and dorsal aspects × 300 (Stein).
10-12. HISTRIO sp. (Sterki).—10, diagrammatic outline of zooid antecedent to subdivision by transverse fission, showing at s_1 and s_2 the newly-growing ventral ciliary series, at pr_2, the second peristomal fringe, and at m_1, m_2, the new elements of the first and second marginal series; 11 and 12, showing disposition and numerical order of development of the newly developing frontal, ventral, and anal series (after Sterki).
13, 14. HISTRIO STEINII, Müll. sp., vol. ii. p. 789.—13, Ventral aspect, × 200; 14, example dividing by transverse fission, with at pr_2 and pr_3, newly developing secondary and tertiary peristomal ciliary systems (Stein).
15-17. STYLONYCHIA PUSTULATA, Ehr., vol. ii. p. 791.—17, Ventral aspect, × 150; 15, zooid disintegrating by diffluence; 16, encysted animalcule (Stein).
18-21. STYLONYCHIA MYTILUS, Ehr., vol. ii. p. 790.—18 and 19, ventral and lateral aspects, × 150; 20, zooid preparing to divide by transverse fission, the rudimentary second peristomal fringe *pr*, being already developed; 21, variety having the frontal, ventral, and anal styles and setæ distinctly fimbriated (Stein).
22. STYLONYCHIA FISSISETA, C. & L., vol. ii. p. 791.— Ventral aspect, × 300 (Claparède & Lachmann).
23, 24. ASPIDISCA LYNCASTER, Stein, vol. ii. p. 793.—23, Ventral and, 24, dorsal aspects, × 250 (Stein).
25-29. ASPIDISCA COSTATA, Duj. sp., vol. ii. p. 794.—25 and 26, ventral and dorsal aspects, × 300 (Stein); 27-29, developmental phases as observed by the author and described at p. 794.
30. GLAUCOMA MARGARITACEUM, Ehr. sp., vol. ii. p. 796, × 150 (Clap. & Lach.)
31-33. ASPIDISCA TURRITA, C. & L., vol. ii. p. 793.—Ventral, dorsal, and lateral aspects, × 229 (Stein).
34-36. URONYCHIA TRANSFUGA, Müll. sp., vol. ii. p. 797.—34 and 35, Dorsal and ventral aspects; 36, example with fimbriated anal uncini, × 250 (Stein).
37. MICROTHORAX SULCATUS, Eng., vol. ii. p. 796, × 220 (Engelmann).
38-40. GLAUCOMA SCINTILLANS, Ehr., vol. ii. p. 795.—38 and 39, Ventral aspect, with at 39, vibratile membrane extended, × 150; 40, encysted zooid dividing by oblique fission (Stein).

PLATE XLV

PLATE XLVI.

EXPLANATION.

FIG.
1, 2. RHYNCHETA CYCLOPUM, Zenker, vol. ii. p. 806.—1, Adult animalcule, × 150; 2, extremity of single tubular sucker, × 600 (Zenker).

3-5. SPHÆROPHRYA UROSTYLÆ, Maupas, vol. ii. p. 809.—3, Normal adult zooid; 4, example dividing by transverse fission, the anterior moiety with temporarily developed cilia; 5, a zooid elongated and with cilia developed preparatory to subdivision, × 200 (Stein).

6. SPHÆROPHRYA PUSILLA, C. & L., vol. ii. p. 808, × 150 (Claparède & Lachmann).

7-9. SPHÆROPHRYA STENTOREA, Maupas, vol. ii. p. 808, × 200 (Stein).

10, 11. TRICHOPHRYA DIGITATA, Stein sp., vol. ii. p. 812.—10, Adult animalcule, × 300; 11, non-tentaculate germ (Stein).

12, 13. TRICHOPHRYA EPISTYLIDIS, C. & L., vol. ii. p. 811.—12, Dorsal; 13, lateral aspects, × 150 (Clap. & Lach.).

14-17. PODOPHRYA ASTACI, C. & L., vol. ii. p. 819.—14, Adult animalcule, × 200; 15 and 16, ciliated embryos; 17, earliest phase of fixed condition (Stein).

18. PODOPHRYA QUADRIPARTITA, C. & L., vol. ii. p. 820.—Two examples attached to stalk of *Epistylis plicatilis*, × 150 (Stein).

19-22. PODOPHRYA FERRUM-EQUINUM, Ehr. sp., vol. ii. p. 813.—19 and 20, Front and profile views showing compressed form of body and proportionate size of pedicle, × 100 (after Clap. & Lach.); 21, ciliated embryo; 22, earliest fixed condition possessing only a single tentacle as represented by Zenker.

23. PODOPHRYA CYCLOPUM, C. & L., vol. ii. p. 818, × 150.

24-30. PODOPHRYA FIXA, Müll. sp., vol. ii. p. 813.—24, Adult animalcule containing at *a* a ciliated embryo, × 200; 25, an example dividing by transverse fission; 26, conjugation of two adjacent animalcules; 27, animalcule forming a membranous encystment (Stein); 28, encystment as found within body of *Stylonychia mytilus* (Engelmann); 29 and 30, ciliated embryos.

31. PODOPHRYA WRZESNIOWSKII, S. K., Stein sp., vol. ii. p. 817, × 150 (Wrzesniowski).

32-35. ACINETA STELLATA, S. K., vol. ii. p. 838.—32, Adult animalcule, × 1000; 33 and 34, encysted examples, the body contents in the first instance divided into two equal halves; 35, free-swimming Sphærophrya-like embryo.

36-39. ACINETA LINGUIFERA, C. & L., vol. ii. p. 831.—36 and 37, Adult zooids, front and profile views, × 150; 38, empty lorica; 39, ciliated embryo (Stein).

EXPLANATION OF PLATE XLVI. (*continued*).

FIG.
40-43. ACINETA MYSTACINA, Ehr., vol. ii. p. 834.—40 and 41, Long and short pedicled varieties, the former with, at *a*, a germinal bud × 150; 42, lorica with contained animalcule seen from above; 43, a germinal bud with contained ciliated embryo further enlarged.

44. ACINETA NOTONECTÆ, C. & L., vol. ii. p. 833, × 150 (Clap. & Lach.).

45-47. ACINETA PATULA, C. & L., vol. ii. p. 835.—45 and 46, Animalcules with tentacles extended and contracted, × 100; 47, example with elongate lorica giving birth to a ciliated embryo (Clap. & Lach.).

48-51. HEMIOPHRYA GEMMIPARA, Hertwig sp., vol. ii. p. 823.—48, Distal extremity of pedicle bearing animalcule with extended tentacles of two orders, suctorial and prehensile, × 150; 49, prehensile rib tentacles more highly magnified, showing spirally disposed granular external sheath; 50, example with six terminal buds, into each of which is produced a prolongation of the branching endoplast, the tentacles not represented; 51, an hypotrichously ciliated embryo, × 400 (Hertwig).

52. SOLENOPHRYA CRASSA, C. & L., vol. ii. p. 828, × 150 (Clap. & Lach.).

53-56. PODOPHRYA MOLLIS, S. K., vol. ii. p. 821.—53 and 54, Adult animalcules, × 150; 55 and 56, successive phases of transverse fission of the same zooid, first manifested by the extension and fixture of two ordinary suctorial tentacles to a more remote point, these two ultimately fusing together and joining the pedicle of the newly produced animalcule.

57. ACINETA MYSTACINA, vol. ii. p. 834.—Ciliated embryo, × 300 (Stein).

58, 59. PODOPHRYA STEINII, C. & L., vol. ii. p. 815.—58, Adult zooid, × 150 (Stein); 59, ciliated embryo, × 300 (Stein).

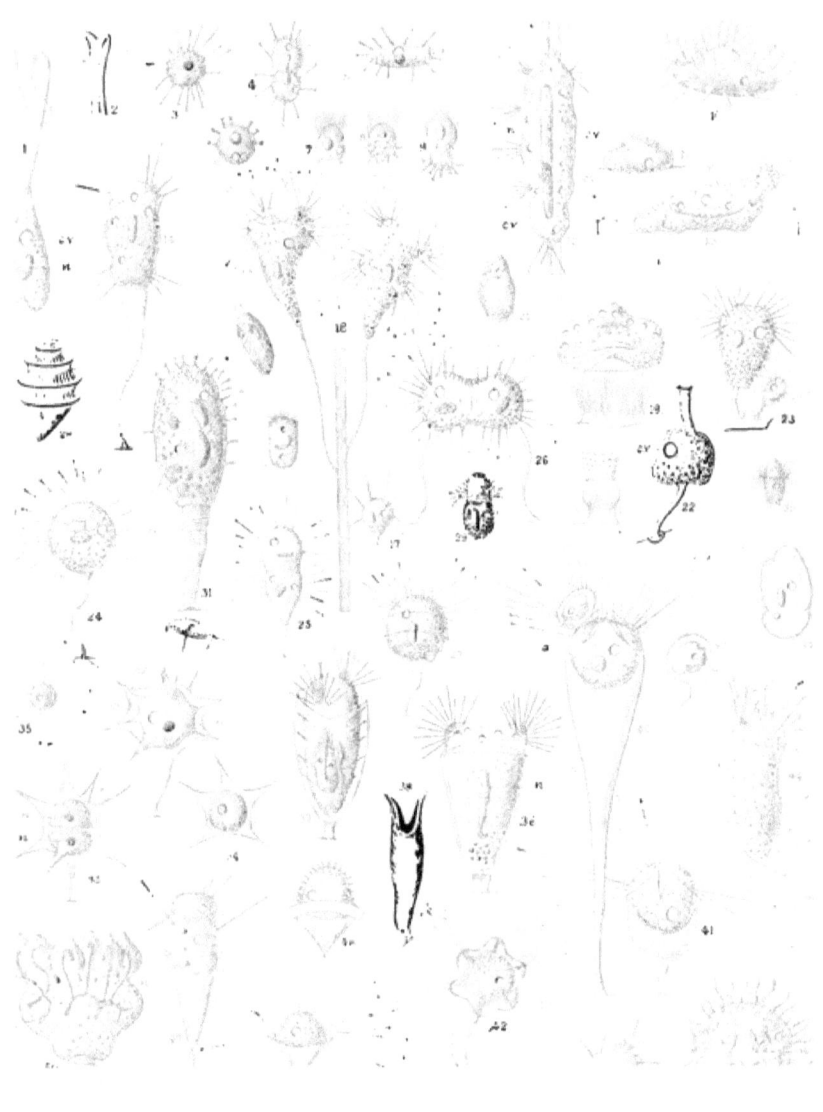

PLATE XLVII.

EXPLANATION OF PLATE XLVII.

FIG.
- 1–5. PODOCYATHUS DIADEMA, S. K., vol. ii. p. 827.—1, Adult zooid, with tentacles fully extended, × 300; 2, zooid with tentacles retracted ; 3, zooid of abnormally minute size as compared with that of the lorica, being probably the relict of a recent fissive process ; 4 and 5, successive developmental forms.
- 6, 7. SPHÆROPHRYA SOL, Mecz., vol. ii. p. 810.—6, Normal zooid, × 250 ; 7, elongate shape assumed preparatory to division by transverse fission (Mecznikow.).
- 8. HEMIOPHRYA CRUSTACEORUM, Haller sp., vol. ii. p. 826.—Zooid with numerous ovate gemmules and distal end of pedicle, × 300 (Haller).
- 9–14. HEMIOPHRYA GEMMIPARA, Htwg. sp., vol. ii. p. 823.—9 and 10, Distal end of pedicle, with zooid at 9 bearing as many as eight elongate ciliated gemmules, and at 10 with two subsphæroidal tentaculiferous embryos, × 200 ; 11–14, phases of development from a free-swimming hypotrichously ciliated embryo towards the parent form (Robin).
- 15. HEMIOPHRYA BENEDENI, Fraip. sp., vol. ii. p. 824.—Zooid with extended tentacles and distal region of pedicle, × 200 (Fraipont).
- 16–22. DENDROSOMA RADIANS, Ehr., vol. ii. p. 841.—16, Erect, slender zoocaulon of young colony-stock, having but three tentaculiferous ramuscules, as delineated by Mr. Thomas Bolton, × 150 ; 17, luxuriantly developed colony-stock as supplied to the author by Mr. Bolton, the decumbent stolon, *st*, giving origin to numerous erect, variously branching zoocaula, some of these, as at *a*, enclosing internally developed ciliated embryos, and others, as at *b b*, with more minute externally developed reproductive capsules, × 50 ; 18, distal region of zoocaula with exogenous germs, that at *a* having short capitate tentacles, × 150 ; 19, portion of zoocaulon with enclosed ciliated embryo *e*, and a portion of the cord-like endoplast, × 400 ; 20, free-swimming ciliated embryo, × 600 ; 21, earliest adherent condition of embryo, the cilia being absorbed and short capitate tentacles developed in their place ; 22, more advanced growth of the same embryo, a single short tentaculiferous prolongation being developed at one extremity, × 600.

PLATE XLVII

PLATE XLVIII.

EXPLANATION.

FIG.
1-4. URNULA EPISTYLIDIS, C. & L., vol. ii. p. 807.—1, Normal zooid with extended tentacula, × 200; 2, zooid in a resting or encysted state, its body-substance having subdivided into four germinal fragments; 3, example with contained ciliated embryo; 4, isolated ciliated embryo, × 300 (Clap. & Lach.).

5. PODOPHRYA LIMBATA, Maupas, vol. ii. p. 816.—Typical zooid with peripheral gelatinous film, × 200 (Maupas).

6, 7. SPHÆROPHRYA MAGNA, Maupas, vol. ii. p. 808.—6, Zooid having seized and in the act of devouring by suction half a dozen examples of *Colpoda parvifrons*, × 300; 7, two tentacles, that at *a* with a central hyaline vacuosity, and that at *b* having the substance of its terminal sucker lacerated through the temporary adhesion and subsequent escape of a ciliated infusorium, × 1280 (Maupas).

8-11a. ACINETA DIVISA, Fraipont, vol. ii. p. 836.—8, Normal zooid with, at *a*, a single anteriorly developed, pyriform germ-capsule, from which a ciliated embryo is in the act of emerging, × 300; 9, germ-capsule with, at *op*, operculum-like differentiation through which the embryo ultimately emerges; 10, ciliated embryo, × 600; 11, abnormally elongated zooid with central band-like endoplast; 11a, empty lorica with, at *s*, platform-like horizontal septum upon which the body of the animalcule ordinarily reposes (Fraipont).

12. ACINETA LIVADIANA, Meresch., vol. ii. p. 828.—Adult zooid, × 250, as observed by the author.

13-17. ACINETA FŒTIDA, Maupas, vol. ii. p. 832.—13, Normal adult animalcule, × 580; 14, lorica with tentacular fascicles as seen from above; 15 and 16, ciliated embryos; 17, young zooid with imperfectly developed lorica (Maupas).

18. ACINETA EMACIATA, Maupas, vol. ii. p. 837, × 300 (Maupas).

19. HEMIOPHRYA THOULETI, Maupas, vol. ii. p. 826, × 200 (Maupas).

20. PODOPHRYA CYLINDRICA, Perty, vol. ii. p. 814, × 250 (Mereschkowsky).

21, 22. PODOPHRYA ELONGATA, C. & L., vol. ii. p. 820.—21, Normal zooid, with at *a a a* tentacles partially retracted and exhibiting a spirally convolute aspect; × 150; 22, one such tentacle as observed by the author, × 800, showing that the spiral aspect is due to the presence of a superficial spirally developed granular crest or film.

23. PODOPHRYA CARCHESII, C. & L., vol. ii. p. 818, × 200.

EXPLANATION OF PLATE XLVIII. (continued).

FIG.
24. ACINETA GRANDIS, S. K., vol. ii. p. 831.—Adult zooid, × 100; affixed near the base of this species, see letter *a*, is delineated an ordinary example of *Acineta lemnarum* found associated with it, under a similar magnification, for the purpose of indicating the relative proportions of these two forms.

25–28. ACINETA TUBEROSA, Ehr., vol. ii. p. 829.—25, Long-stalked variety, × 400; 26, example as seen in horizontal optical section showing the four points of attachment of the body to the lorica and the bases of the tentacular fascicles; 27, a similar body as seen in vertical section; 28, horizontal optical section immediately above the junction of the pedicle with the lorica (Fraipont).

29. ACINETA SAIFULÆ, Meresch., vol. ii. p. 836.—Spirit-preserved example, × 200 (Mereschkowsky).

30, 31. HEMIOPHRYA BENEDENI, Fraipont, vol. ii. p. 824.—30, Zooid, × 70; 31, portion of pedicle, × 300, showing quadrangular contour and transversely striated central core.

32, 33. ACINETA CRENATA, Fraipont, vol. ii. p. 837.—Zooid with elongate lorica, × 600; 33, shorter lorica, with distal region only of the pedicle (Fraipont).

34, 35. ACINETA JOLYI, Maupas, vol. ii. p. 835.—34, Front and, 35, lateral view, × 200 (Maupas).

36, 37. OPHRYODENDRON BELGICUM, Fraipont, vol. ii. p. 853.—36, Vermiform and, 37, proboscidiform zooids, × 400 (Fraipont).

38–40. OPHRYODENDRON SERTULARIÆ, Str. Wright sp., vol. ii. p. 851.—38, Ordinary proboscidiform zooid in lateral view, bearing at *v* a single vermiform germ, × 200; 39, proboscidiform zooid from above, showing, after treatment by the author with osmic acid, orbicular contour of body and enclosed branching endoplast; 40, detached vermiform zooid more highly magnified, showing acetabular character of the basal region.

42. ACINETA VORTICELLOIDES, Fraipont, vol. ii. p. 837.—Distal region showing rudimentary development of the lorica, × 300 (Fraipont).

43–45. ACINETOPSIS RARA, Robin, vol. ii. p. 855.—43, zooid with extended tentacle, × 200; 44 and 45, tentacle in various states of contraction more highly magnified and showing externally developed spiral fibrilla (Robin).

46. HEMIOPHRYA TRUNCATA, Fraip., vol. ii. p. 825.—Adult zooid, × 250 (Fraipont).

PLATE XLVIII.

PLATE XLVIIIa.

EXPLANATION OF PLATE XLVIIIa.

FIG.
1-3. EPHELOTA CORONATA, Str. Wright, vol. ii. p. 846.—1, 2, Animalcules with tentacles extended, at *m m* captured monads, × 200 ; 3, tentacles retracted.
4. PODOPHRYA CONIPES, Meresch., vol. ii. p. 815, × 250 (Mereschkowsky).
5. EPHELOTA TROLD, C. & L., vol. ii. p. 847, × 175 (Claparède & Lachmann).
6. ACTINOCYATHUS CIDARIS, S. K., vol. ii. p. 848, × 300.
7. ACINETA TUBEROSA, Ehr., vol. ii. p. 829.—Two zooids with tentacles in a fully extended and partially retracted state, × 200.
8-12. DENDROCOMETES PARADOXUS, Stein, vol. ii. p. 839.—8, Adult animalcule, × 400, one of the branched tentacles at *a* grasping a captured monad ; 9, the same tentacle, × 1200, and showing the tubular perforations of the distal terminations ; 10, a young recently attached animalcule; 11, animalcule enclosing a ciliated embryo ; 12, a detached ciliated embryo (Wrzesniowski).
13-15. OPHRYODENDRON ABIETINUM, C. & L., vol. ii. p. 850.—13, Proboscidiform zooid with proboscis retracted and with vermiform zooid attached to one side, × 100; 14, proboscidiform zooid with proboscis extended ; 15, ciliated embryo, × 200 (Clap. & Lach.).
16-19. OPHRYODENDRON PEDICELLATUM, Hincks, vol. ii. p. 852.—16 and 17, Proboscidiform zooids with proboscis extended and retracted ; 18 and 19, two vermiform zooids, the one in the latter instance having a second example sessilely attached to its body, dimensions unrecorded.
20-25. OPHRYODENDRON PORCELLANUM, S. K., vol. ii. p. 852.—20, Zooid with proboscidiform appendage extended and recurved, × 200 ; 21, the same appendage more completely extended in a rectilinear direction ; 22, a zooid bearing both a proboscidiform and vermiform appendages ; 23, an example having only a vermiform appendage ; 24, distal extremity of proboscis highly magnified, × 800 ; 25, distal region of shaft of the proboscis, × 1200, showing its transverse corrugation.
26-31. OPHRYODENDRON MULTICAPITATUM, S. K., vol. ii. p. 854.—26, Sessile zooid bearing four proboscidiform organs, × 300 ; 27, sessile example with four proboscidiform organs *pr*, and three ovate gemmules *s* ; 28, stalked example with two proboscidiform organs, two gemmules, and one vermiform appendage ; 29-31, young pedicellate zooids ; 30 and 31, bearing each a single proboscidiform organ only, this structure in the latter instance being retracted ; 29, representing a still younger zooid entirely destitute of appendages, and, except for the presence of a pedicle, corresponding with the gemmular bodies represented in figures 27 and 28.
32. OPHRYODENDRON ABIETINUM, C. & L., vol. ii. p. 850.—Vermiform larva with basal chitinous rod, × 400 (after Robin).

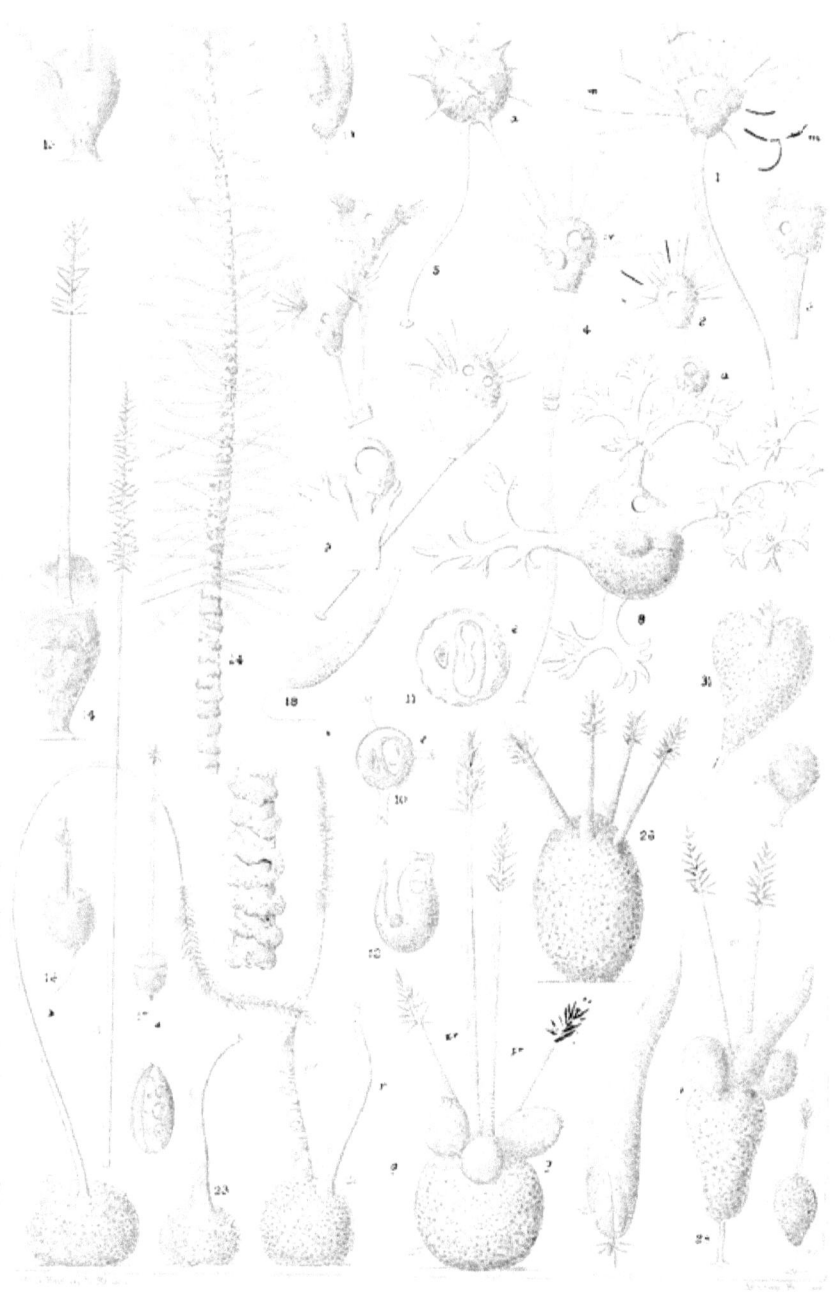

PLATE XLIX.

EXPLANATION OF PLATE XLIX.

Constituting a key to the numerous species of the genus *Vorticella*, each specific form being represented in diagrammatic outline at its most typical condition of extension and in accordance with the scheme suggested in vol. ii. p. 672. To facilitate identification the specific name is appended to each type.

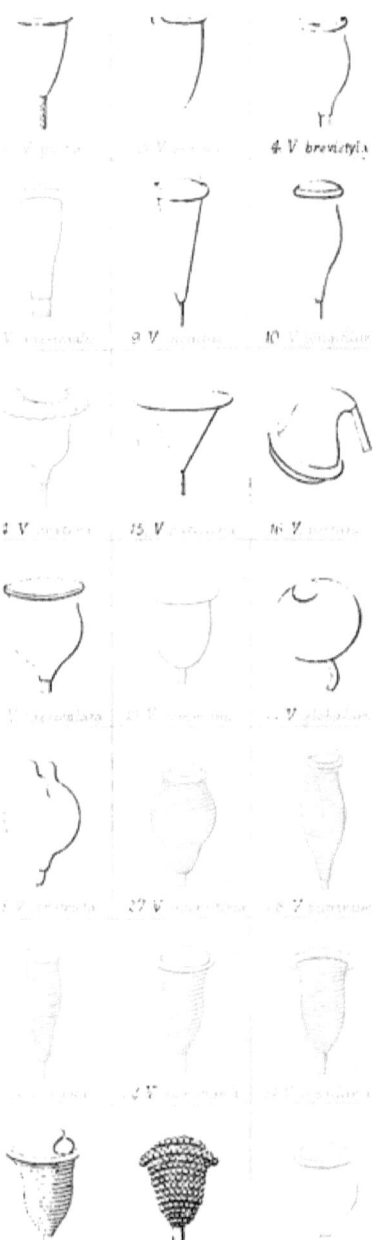

PLATE L.

Illustrating the more important modifications of the nucleus or endoplast and nucleolus or endoplastule as developed among the Infusoria.

EXPLANATION.

FIG.
1. Simple spheroidal nucleus or endoplast, with enclosed nucleolus or endoplastule, of a collared monad, *Codosiga*, the same type being distinctive of the majority of the Flagellata.
2. A single germinal fragment or spore, with contained endoplast and endoplastule, of *Dictyocysta cassis*, × 1000 (Haeckel).
3-5. Modifications of the endoplast of *Euglena viridis*.—3, representing the normal form; 4 and 5, progressive phases resulting in the subdivision of its entire mass into sporular elements × 600 (Stein).
6. Fusiform endoplast, with centrally enclosed endoplastule, of *Chilodon cucullulus*, × 480 (Wrzesniowski).
7-9. Various modifications of the endoplast of *Leptodiscus medusoides* (Hertwig).
10. Ovate endoplast having a reticulate granular consistence, with laterally attached endoplastule, of *Acineta fœtida*, × 1280 (Maupas).
11. Subspheroidal endoplast, with numerous enclosed endoplastules, of *Acineta Jolyi*, × 900 (Maupas).
12. Spheroidal granular endoplast of Rhizopod *Pelomyxa villosa*, × 1000 (Leidy).
13. Twin endoplasts, with connecting cord or funiculus, of *Litonotus fasciola*, × 480 (Wrzesniowski).
14. Endoplastic system of *Loxodes rostrum*, consisting of numerous spheroidal endoplasts with both enclosed and externally developed endoplastules connected together by a thread-like funiculus. In some instances, *a a*, the external endoplastules are attached to the funiculus, × 600, and treated with acetic acid and iodine (Wrzesniowski).
15. Twin endoplasts with connecting funiculus of *Litonotus diaphanus* after treatment with acetic acid and iodine, × 600 (Wrzesniowski).
16. Endoplast of *Chilodon cucullulus* with both an internally developed and laterally attached endoplastule, × 400 (Wrzesniowski).
17. Twin endoplasts, with laterally attached endoplastules, of *Stichotricha aculeata*, × 600, + acetic acid (Wrzesniowski).
18. Endoplastule of *Pleuronema chrysalis* after treatment with acetic acid (Bütschli).
19, 20. Modifications of the endoplast of *Nyctotherus cordiformis*.—19, normal form, with laterally attached endoplastule; 20, endoplast become greatly enlarged and metamorphosed into an elaborately convoluted coil (Stein).
21. Band-like endoplast with *a*, investing membrane and *b*, *b*, contained endoplastules of *Vorticella microstoma*, × 450 (Stein).
22, 23. Band-like endoplasts of *Carchesium polypinum*, subsequent to conjugation; at 22, earlier phase with numerous enclosed endoplastules; at 23, more advanced stage in which the internal substance has become separated into as many as ten germinal masses which are held together only by the delicate bounding membrane of the originally continuous and homogeneous endoplast, × 1000 (Greef).

EXPLANATION OF PLATE L. (continued).

FIG.
24. Conjoint endoplasts of *Amphileptus anas* after treatment with acetic acid, their substance being thus shown to be divided up into numerous polygonal fragments, each with a minute central refringent corpuscle, × 600 (Bütschli).
25-28. Endoplastules or nucleoli of *Paramœcium putrinum;* 25 and 26, normal forms, × 2000 + acetic acid ; 27 and 28, successive phases during the process of subdivision or fission (Bütschli).
29-31. Endoplastules of *Carchesium polypinum;* 29, living condition ; and 30, after treatment with acetic acid, × 1500 ; 31, example elongated preparatory to subdivision (Bütschli).
32. Endoplast of *Paramœcium bursaria*, during process of subdivision, the normal, single, laterally attached endoplastule having already separated into two, × 800 (Bütschli).
33. Twin endoplasts of *Stylonychia mytilus;* aspect exhibited at the commencement of the fissive process, the endoplastules in the vicinity of each endoplast also commencing to subdivide, × 600 (Bütschli).
34. Twin endoplasts of *Stylonychia mytilus* in a more advanced stage of subdivision, the endoplasts having already separated into four, both these and the endoplasts presenting a striated aspect, × 500 (Bütschli).
35-37. Variously branched endoplast of *Ophrydendron belgicum;* 35 and 36 derived from proboscidiform, and 37 from a vermiform zooid, × 800 (Fraipont).
38. Ramifying endoplast from an erect main trunk of *Dendrosoma radians*, as observed by the author, × 800.
39. Endoplastic system of *Loxophyllum meleagris*, consisting of numerous irregularly ovate nodular endoplasts, united to each other by a delicate thread-like funiculus, × 600, + acetic acid, iodine, and picrocarmine (Bütschli).
40. Irregular nodular endoplast of *Hemiophrya Thouleti*, × 500 (Fraipont).
41, 42. Moniliform endoplast of *Stentor polymorphus;* 41, the entire structure, × 200 ; 42, a fragment more highly magnified showing internal refringent corpuscles and connecting funiculus (Stein).
43. Anterior region of undulating band-like endoplast of *Stentor Roselii*, from the extremity of which one germ-sphere or embryo, with a central endoplastule (the future endoplast) and incipient contractile vesicle, has become already constricted off, a second similar germ-sphere being in an advanced state of development, × 500 (Claparède and Lachmann).
44. Twin endoplasts of *Stylonychia mytilus*, normal aspect, showing a distinct investing membrane, connecting funiculus, and laterally attached endoplastules, × 600 (Bütschli).
45, 46. Single endoplastic elements, with attached endoplastules, of *Stylonychia mytilus*. —45, Showing distinct differentiation of the anterior and posterior moieties ; 46, exhibiting at one extremity the delicate investing membrane, × 400 (Stein).
47. More abnormal elongated endoplast of *Stylonychia mytilus* showing distinct bounding membrane and three striated laterally located endoplastules, × 400 (Stein).
48. Single endoplast, with striated laterally attached endoplastule, of *Kerona polyporum*, × 300 (Stein).
49. Single endoplast of *Oxytricha* sp., whose substance has divided up into numerous polygonal germs, × 600 (Bütschli).
50. Single endoplast, with contained germ-spheres, of *Urostyla grandis*, × 600 (Bütschli).
51. Compound racemose endoplast of *Plagiotoma lumbrici*, × 400 (Stein).
52. Branching endoplast of young zooid of *Dendrosoma radians* as observed by the author, × 800, + osmic acid and picrocarmine.
53, 54. Branching, convoluted endoplast of *Acineta mystacina*, × 600 (Fraipont).

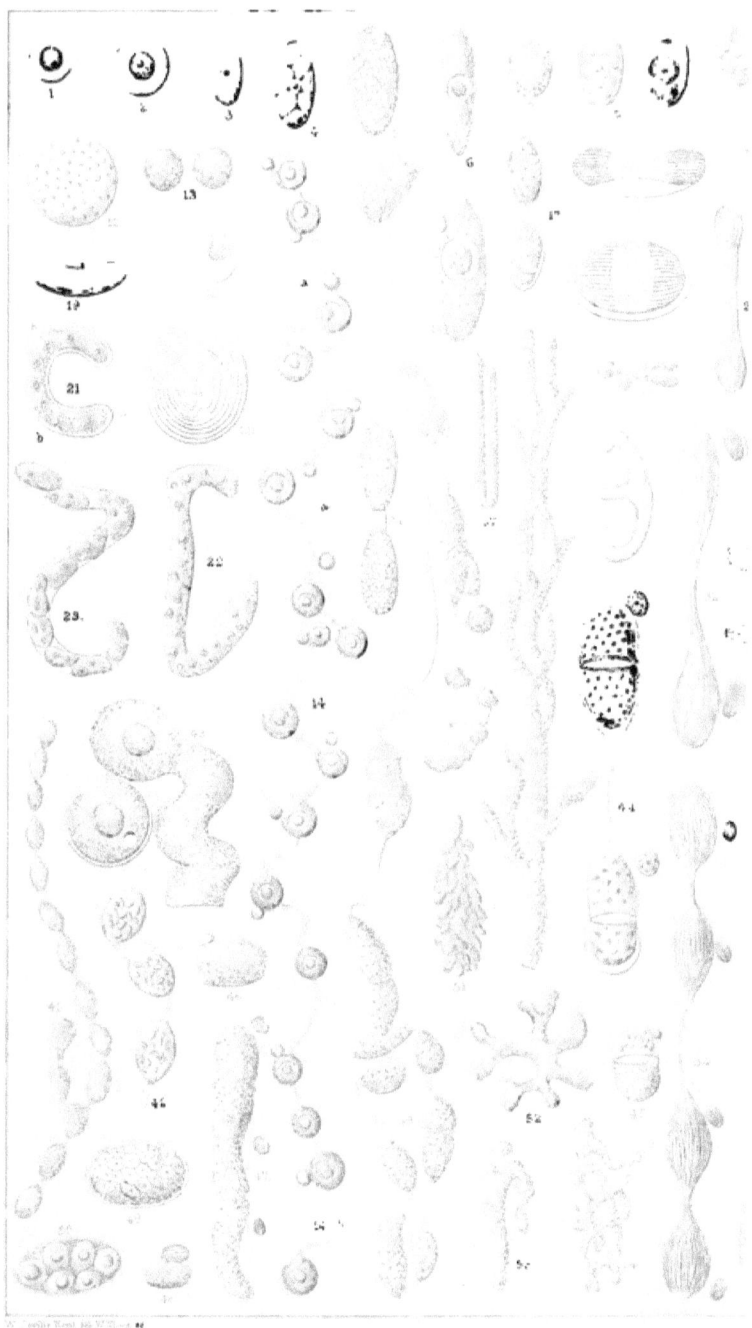

PLATE LI.

EXPLANATION.

FIG.

1-5. Illustrating the apparatus employed by Messrs. Dallinger and Drysdale during their prolonged investigation of the life-histories of various monads referred to at vol. i. p. 116, originally figured and described in the 'Monthly Microscopical Journal for March 1874.—1, $a\,a$, glass plate adapted to fit the stage of the microscope ; b, circular aperture cut in the plate, a thin piece of glass c, d, e, f being cemented over it to permit the near approach of the achromatic condenser ; $g\,g\,g$, brass socket with ring attached which is fixed with marine glue to the projecting arm of the glass plate a, and supports a cylindrical reservoir of water, Fig. 4 ; $h\,h\,h\,h$, outline of position of bibulous paper having central aperture and tongue-like projection which dips down into the reservoir fitted to the socket g ; 2, bibulous paper cut to fit the glass stage, the portion $b\,c$ leading into the reservoir g of Fig. 1 ; 3, moist chamber consisting of a short piece of glass tubing a, having the bottom edge e carefully ground, the top having over it a thin elastic film with a minute central perforation c, and securely fastened to the sides at the groove d ; 4, cylindrical glass reservoir fitting into the socket g of Fig. 1, and into which the projecting arm of the bibulous paper dips ; 5, the entire apparatus in working order, the object glass g being racked down through the central perforation of the elastic film f of the moist chamber ch ; $a\,a$ section of glass stage ; b, aperture in the same ; c, the glass cemented over the aperture ; d, covering glass over object examined ; e, walls of moist chamber.[1]

6. Chamber invented by Professor Tyndall, referred to at vol. i. p. 130, originally figured and described in the 'Transactions of the Royal Society,' 1877, for the perfect isolation and cultivation of organic infusions and equally suited for a similar culture of Infusoria. c, central box or chamber, the front being removed showing the windows w, w, for the admission of light ; t, six test-tubes fitting with an air-tight packing into the floor of the chamber ; a, b, sinuous glass tubing permitting the access of air but not of germs to the chamber ; p, pipette with stuffed funnel, fitting into a pin-hole perforation in a piece of indiarubber and stuffing-box containing cotton-wool moistened with glycerine, and thus permitting its insertion and withdrawal without the introduction of adventitious germs.

EXPLANATION OF PLATE LI. *(continued)*.

FIG.
7, 8. Illustrating the arrangement of the microscope and lamp employed by the author for obtaining the most satisfactory illumination and definition of minute flagellate organisms, when working with object-glasses of 1–16th to 1–50th inch nominal focal distance, for which he is chiefly indebted to a most kind and painstaking demonstration by Mr. E. M. Nelson, F.R.M.S. The mirror *m* being turned to one side, the microscope and lamp are so disposed that the central ray of light *ax* from the *narrow edge* of the lamp flame passes through the optical axis of the achromatic condenser *a c*, and is then focussed upon the field of view, by means of the substage rackwork, in such a manner that employing a 1-inch object-glass, a sharply defined image of the lamp-flame, edge on, is projected upon the centre of the field in company with the objects under examination as shown at Fig. 8. If the 1-inch object-glass is now detached, and a 1–16th, 1–25th or 1–50th substituted, and focussed into place, a slight readjustment of the centering of the achromatic condenser being perhaps required, it will be found that the entire field is brilliantly illuminated, and the most minute objects defined with an amount of sharpness rarely obtained under other conditions. In addition to the ordinary graduating diaphragm placed immediately beneath the lenses of the achromatic condenser as at d^1 in Fig. 7, the author has derived considerable advantage from the interposition of a second diaphragm at d^2, or the lowest point in the substage arrangement.

9, 10. Trichocysts of *Bursaria (Panophrys) leucas* (see vol. i. p. 82), as interpreted by Professor G. J. Allman; Fig. 9, trichocysts, *tr, in situ*, disposed in an even vertical layer immediately beneath the cuticle and locomotive cilia *c*, × 1000. Fig. 10, the same trichocysts projected irregularly from the entire periphery as hairlike filaments or setæ with recurved distal ends, on the application of acetic acid or forcible compression (after Allman, 'Quarterly Journal of Microscopical Science,' vol. iii., 1855).